T0212122

Lecture Notes of the Institute for Computer Sciences, Social Informatics and Telecommunications Engineering 319

More information about this series at http://www.springer.com/series/8197

Rogério Mugnaini (Ed.)

Data and Information in Online Environments

First EAI International Conference, DIONE 2020
Florianópolis, Brazil, March 19–20, 2020
Proceedings

 Springer

Editor
Rogério Mugnaini 🔟
Universidade de São Paulo
São Paulo, Brazil

ISSN 1867-8211 ISSN 1867-822X (electronic)
Lecture Notes of the Institute for Computer Sciences, Social Informatics
and Telecommunications Engineering
ISBN 978-3-030-50071-9 ISBN 978-3-030-50072-6 (eBook)
https://doi.org/10.1007/978-3-030-50072-6

This Springer imprint is published by the registered company Springer Nature Switzerland AG
The registered company address is: Gewerbestrasse 11, 6330 Cham, Switzerland

Preface

It is a pleasure to present the proceedings of the First EAI International Conference on Data and Information in Online environments (DIONE 2020), held in Florianópolis, Brazil, March 19–20, 2020.

The production, dissemination, and evaluation of contents in online environments place new challenges and directions for the metrics of science, such as the data processing tools and techniques to interpret the contents generated in the social media context. At the core of the study of these aspects, lies the intersection of Computer Science, Information Science, and Communication since the study of these topics requires a multidisciplinary view. With the purpose to open an academic, investigative, and practical space, the first edition of DIONE, which promises to become a well-established conference, was launched, discussing cutting-edge topics on network science, data science, data engineering, social media metrics, natural language processing, scientometrics, and big data. Being a new conference, DIONE was able to bring researchers, scholars, faculty members, and doctoral and post-doctoral students together to present and discuss innovative ideas related to the aforementioned topics.

DIONE 2020 consisted of 18 full papers out of 37 submissions; 67% of the papers were from Brazilian authors while the 33% remaining were from other countries including Belgium, Canada, Peru, Portugal, Slovakia, and the USA. We highly appreciate the interest of all the authors who trusted DIONE to submit their research results. Special thanks go to the Organizing Committee, the members of the Technical Program Committee, and the external reviewers for their willingness to collaborate. In this edition, we are privileged to have two keynote speakers: Prof. Miriam Capretz from Western University, Canada, who delighted us with a presentation on opportunities, challenges, and applications of big data; and Prof. Mario Dantas from the Federal University of Juiz de Fora, Brazil, whose lecture addressed the Data-Intensive Scalable Computing (DISC), a data-centric approach containing a wide diversity of applications usually employed in the web and business environments.

We sincerely appreciate the guidance and support provided by EAI. Last but not least, we would like to thank the speakers and attendees for being a part of this conference.

We hope that these proceedings will be of interest for all the sectors involved with the aforementioned topics and that all the particular research results will help to bring more insights in order to continue strengthening these interdisciplinary issues.

April 2020   Rogerio Mugnaini

Organization

Steering Committee

Imrich Chlamtac	Bruno Kressler Institute, University of Trento, Italy
Carlos Luis González-Valiente	European Alliance for Innovation, Slovakia

General Chair

Douglas Dyllon Jeronimo de Macedo	Federal University of Santa Catarina (UFSC), Brazil

Program Chair

Rodrigo Costas	CWTS, Leiden University, The Netherlands

Publicity and Social Media Chair

Orlando Gregorio Chaviano	Pontificia Universidad Javeriana, Colombia

Workshops Chair

Mario Antonio Ribeiro Dantas	Federal University of Santa Catarina (UFSC), Brazil

Publications Chair

Maria Manuel Borges	University of Coimbra, Portugal

Local Chair

Nancy Sánchez-Tarragó	University of Pernambuco, Brazil

Web Chair

María Josefa Peralta González	Central University of Las Villas, Cuba

Conference Manager

Barbora Cintava	European Alliance for Innovation, Slovakia

Technical Program Committee

Ricardo Arencibia Jorge	Empresa de Tecnologías de la Información, Cuba
Stefanie Haustein	University of Ottawa, Canada
Enrique Orduña Malea	Polytechnic University of Valencia, Spain
Diana Lucio Arias	Universidad Javeriana de Colombia, Colombia
Juan Gorraiz	University of Vienna, Austria
Elías Sanz Casado	Carlos III University of Madrid, Spain
Fabrizio Messina	University of Catania, Italy
Alessandra de Benedictis	Università Federico II di Napoli, Italy
Manik Sharma	DAV University Jalandhar, India
Adán Hirales Carbaja	CETYS University, Mexico
Ulises Orozco Rosa	CETYS University, Mexico
Juan Wang	Delta Micro Technology Inc., USA
Wayne Buente	University of Hawaii, USA
Xuan Liu	Future Network Research Center, Southeast University, China
Buğra Gedik	Bilkent University, Turkey
Claudio Silvestri	Ca' Foscari University of Venice, Italy
Joe Tekli	Labanese American University, Lebanon
Lakshmish Ramaswamy	University of Georgia, USA
Li Xiong	Emory University, USA
Carol Tenopir	University of Tennessee, USA

Contents

Data Processing

Use and Analysis of Network Information

Evaluation of Science in Social Networking Environment

Data Processing

Data Management Plan in Research: Characteristics and Development

Paulo A. Cauchick-Miguel[1]([✉]), Suzana R. Moro[1], Roberto Rivera[2], and Marlene Amorim[2]

[1] Universidade Federal de Santa Catarina, Florianópolis, SC 88040-900, Brazil
`paulo.cauchick@ufsc.br`
[2] Universidade de Aveiro, 3810-193 Aveiro, Portugal

Abstract. Data science is an interdisciplinary field that extracts value from data. One of the relevant areas is its application in research in order to define requirements of the data life cycle. Thus, data should be managed before, during, and after a research project completion. A robust data management plan (DMP) is a relevant and useful instrument to establish data-related requirements. In this context, this paper aims at highlighting some characteristics associated to research data management. To conduct this study peer-reviewed literature and secondary data are methodologically employed to fulfil the paper objective. The results discuss the development of DMP, provide some examples of documents and a check list related to data management, and present some recommendations for developing a suitable data management plan from the literature. The data management plan is one of the important instruments that should be considered with care when designing and applying it. Future work may consider providing a structure and guidance for research students in the field of industrial engineering as a valuable avenue to explore.

Keywords: Data planning · Research plan · Research Data Management

1 Introduction

Data science is a broad and an interdisciplinary domain present in various subjects, such as computing, statistics, biology, engineering, and many others. One of the relevant areas is associated to managing data in research in order to specify data requirements from data discovery/collection to dissemination and data conservation [1].

Although researchers usually establish and record data collection and analysis procedures, it is not so usual to address with the same extent of detail other stages of the data life cycle particularly in the field of industrial engineering. To this end, a robust data management plan (DMP) is a relevant and useful instrument to establish data-related requirements. DMP provides structured data-related contents during and after a completion of a research project [1]. Indeed, data management has been a critical skill for researchers [2, 3]. Moreover, in recent years most of funding agencies (e.g. in Europe and

R. Mugnaini (Ed.): DIONE 2020, LNICST 319, pp. 3–14, 2020.
https://doi.org/10.1007/978-3-030-50072-6_1

in the US) require that data management plans must be submitted as part of a research proposition [1, 3].

In this context, the aim of this paper is to highlight some characteristics with regard to data management for research. Another potential outcome of this study is the knowledge of specificities related to data management plan that might help beginners when starting to deal with this kind of instrument.

The remainder of this paper is structured as follows. Section 2 briefly outlines some concepts that support the study development. Section 3 describes the research procedures while Sect. 4 outlines the outcomes and discussion related to the scope of this work. Finally, Sect. 5 draws some concluding remarks from this work.

2 Brief Literature Background on Data Management

Data can be defined as "everything that would be needed to reproduce a given scientific output" [4]. Research data management (RDM) could be described as the process that requires one to create, organize, keep, find, and share data [5]. Data management guarantees that the process used by the researcher to the collection is organized, clear, and understandable [4], aiming to support the research process reproducibility. Data is exposed to risks during the lifecycle of a research project, and it can affect their fair use and interpretation; therefore, data management helps to ensure its availability [6]. Proper data management also becomes necessary to strengthen open science by sharing accurate data [7].

There are many visualizations of RDM as a lifecycle, although no consensus exists [5]. Data management lifecycle encompasses [4]: (i) creating or collecting data; (ii) processing data; (iii) analysing the data; (iv) preserving data; (v) providing access to data; and (vi) reusing the data to conduct new studies. New RDM lifecycle visualizations could be helpful to stakeholders to understand their tasks and responsibilities as well as to explain the logic and order of activities in the research process [5].

Researchers still need guidance on practices that could facilitate data usage and understanding of specialized concepts, for instance, metadata and policies for data use and reuse [8]. Therefore, RDM is still a challenge for researchers and research institutions, that is amplified with the growing role that digital disruption plays in research activities and in the variety of tools available [9]. According to Mannheimer [8], most researchers practice internal data management to prevent losing data, to maintain a constant workflow even with researchers' turnover, and to make sharing within the team easier.

A data management plan is a written description document detailing how a researcher plans to collect, store, describe, preserve, and make data available [10]. The DMP goes through peer review and can be used in part to evaluate a research project merit [1]. Michener [1] adds that plans also help to document data management activities related to funded projects that may be re-examined throughout performance reviews. DMP could vary in length from a short document (from 1 to 2 pages) constructed by the primary investigator to a multi-page report prepared within the frame of a research project to direct the team through all aspects of the research data management [11].

After the US National Science Foundation announced that all funding proposals from 2011 onward must include a DMP, the interest in the topic has grown exponentially

[10, 12]. Other federal funding agencies, such as the National Aeronautics and Space Administration, require DMP and many others are going to require shortly [13]. In the 2014–16 work programme of Horizon 2020, the Open Research Data Pilot (ORD pilot) was included in selected areas; in 2017, the revised version extended ORD pilot to all thematic areas [14]. Many European universities are supporting open scholarship and science and initiatives in the field of open data [15].

Publishers are also encouraging researchers to use RDM. For instance, Elsevier [16] provides free access e-learning resources to support researchers on managing and publishing research data, as well as creating a DMP and benefiting from citing data. Besides, Research Data Management Librarian Academy – RDMLA [17] is a group of university libraries partnering with Elsevier to provide online RDM training.

No unique formula for appropriate data management exists, as the task depends on the science domain, the types of data, and the size of the project [7]. Nevertheless, a DMP normally includes [11]: (i) data that will be created; (ii) data documentation and organization; (iii) data storage and security; (iv) data management and preservation after project completion; and (v) data accessibility for reuse and sharing. Experience and knowledge gained in developing DMP will be valuable over the next years [15].

Van der Loon *et al.* [13] compare DMP from different areas and academic units and highlight that shortcomings vary across them. Many DMP analysed by the authors provides an ambiguous or insufficient description of how research data will be managed, preserved, and shared. The authors also point out that engineering researchers rarely mention data sharing through supplemental material enclosed in journal articles.

There are several tools available to help researchers develop the skills needed to make choices and guide users to create data management plans [10]. Some outlets analysed the usage of some of them, such as: ERDMAS [9], MANTRA [10], DMP tool [11, 12], and DMP online and IEDA Data Management Plan Tool [12].

Reilly and Dryden [12] recommend that each institution should build its own tool to comprise specific local information concerning their researcher funding activities, including resources, services available on campus, particular policies, and requirements. Therefore, there is a need that research agencies get engaged with distinct scientific communities to develop data management plan formats that best suit specific disciplines [7]. RDM service providers also need to build the expertise to address the needs of individual researchers or research groups [15].

2.1 FAIR Data Principles

DMP requirements from funding agencies are generally committed with standards like the FAIR Data Principles [8]. For instance, European Commission Horizon 2020 FAIR DMP [14] provides such requirements. In general terms, FAIR data principles [18] states that data must be structured to be "Findable", "Accessible", "Interoperable", and "Reusable". Figure 1 illustrates FAIR guiding principles.

2.2 Data Life Cycle in Research

Figure 2 depicts the relationship between hypothetical research and data life cycle. As can be seen in Fig. 2, there are a number of stages in which data is present before, during,

To be **F**indable	F1. (meta)data are assigned a globally unique and persistent identifier
	F2. data are described with rich metadata (defined by R1 below)
	F3. metadata clearly and explicitly include the identifier of the data it describes
	F4. (meta)data are registered or indexed in a searchable resource
To be **A**ccessible	A1. (meta)data are retrievable by their identifier using a standardized communications protocol
	A1.1 the protocol is open, free, and universally implementable
	A1.2 the protocol allows for an authentication and authorization procedure, where necessary
	A2. metadata are accessible, even when the data are no longer available
To be **I**nteroperable	I1. (meta)data use a formal, accessible, shared, and broadly applicable language for knowledge representation
	I2. (meta)data use vocabularies that follow FAIR principles
	I3. (meta)data include qualified references to other (meta)data
To be **R**eusable	R1. meta(data) are richly described with a plurality of accurate and relevant attributes
	R1.1. (meta)data are released with a clear and accessible data usage license
	R1.2. (meta)data are associated with detailed provenance
	R1.3. (meta)data meet domain-relevant community standards

Fig. 1. FAIR guiding principles [18].

and after finishing a research project. The circles with numbers mean details that are linked to the data life stages. The researchers usually: (1) test hypotheses; (2) collect data; (3) interpret data; (4) disseminate by publishing. Considering the data life stages at 'B', researchers normally: (1) develop a data plan; (2) discover/collect and (3) organize data; (4) ensure data quality; (5) describe and (6) analyse the data; and (7) preserve and (8) share the data with other researchers.

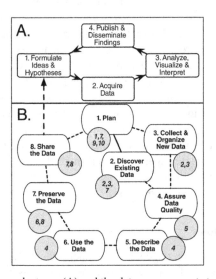

Fig. 2. Research stages (A) and the data management stages (B) [1].

A data plan should cover all or portions of the data life cycle [1]: from data discovery, collection, and organization (e.g., spreadsheets, databases), through quality assurance and control, documentation (e.g., data types, research methods and others) and data usage, to data preservation and sharing with others (e.g., data policies and dissemination approaches). Michener [1] also suggests ten points that can guide the process of creating an effective data management plan for managing research data. After presenting this brief literature section, research procedures are outlined next.

3   Research Procedures

In order to carry out this study, articles in peer-reviewed literature (journals and selected conference papers) were considered in addition to non peer-reviewed material and examples being published by practitioners. At this stage, this collection was not comprehensive but enough for building a brief theoretical background mostly centered in data management as well as data management plan.

3.1   Peer-Reviewed Literature

A literature search was performed to identify relevant publications on data management subject, following recommended research procedures elsewhere [19]. Scopus and Web of Science were accessed to carry out the search, as they encompass a large number of publications from different areas. The key-words "Data management plan*" OR "research data management" were used searched in publication's title. No data filter was used. The results obtained in the databases were respectively 214 and 165 articles at Scopus and Web of Science data bases.

As this is an initial phase of an investigation, the focus was on the most cited articles as a criterion for relevance, using the databases' selection options. A retrospective procedure [19] was also employed to include relevant publications cited by the selected articles. Considering this 'fast-track' process, 24 documents were considered with the aim at constructing a brief literature background on data management and DMP, as mentioned earlier in this section and presented at previous section. This paper does not provide results from the literature analysis of those 24 documents in the results section (this is working in process). In addition, the literature search and analysis is going to be expanded in the near future.

3.2   Other Sources - Non Peer-Reviewed Material

This work also used secondary data from institutions, in the UK (Digital Curation Centre), in Portugal (University of Aveiro), and the European Commission. The Digital Curation Centre (DCC) has been widely cited in the literature, and they are very influential at the practice level as well. As many other universities in Europe and in the USA, University of Aveiro has provided information throughout various sources, seminars, reports, blogs, and webinars on data management and the development of data management plans in the context of the Open Research Data Pilot in Horizon 2020 from the European Commission [20].

3.3   Work Purpose

The objective of this work is to highlight some characteristics related to the life cycle of data management for research, in addition to some examples provided either from some academic practices (e.g. internal documents associated with data management life cycle) or from the peer-reviewed literature (e.g. recommendation of points for constructing a DMP by ref. [1]). The authors of this paper also understand that the knowledge related to data management plan (DMP) might help beginners when starting to deal with this subject.

4 Results: Some Examples of Application

4.1 Developing a Data Management Plan

Digital Curation Centre [21] provides guidance and examples for building a Data Management Plan. There are contents that researchers typically expect to cover in a DMP, related to the following categories for data management [22]: (i) administrative data; (ii) data collection; (iii) documentation and metadata; (iv) ethical and legal compliance; (v) storage and backup; (vi) selection and preservation; (vii) data sharing; and (viii) responsibilities and resources. Table 1 provides a summary of those categories and summarises its contents.

Example of Documents of 'Administrative Data' in a University. Currently,   the University of Aveiro participates in 433 national projects and 95 international projects [23]. As the university progress towards its goals, gradually it creates enormous amount of data that need to be managed. This instance considers research the data organization process used in an international project (Erasmus+Programme). One of the goals of this research project is to contribute to the Europe 2020 Strategy for growth, jobs, social equity, and inclusion, as well as the aims of the strategic framework for European cooperation in education and training - ET 2020 [24].

The identification data of the proposed ideas and participating institutions are depicted in Fig. 3, which shows two forms for data recording. This example illustrates the initial process of the program, in which several European organizations, mainly linked to education and research, come together to discuss ideas and define joint projects to be supported by the Erasmus+programme.

The forms in Fig. 3 provide basic data before research project initiation within the scope of 'Administrative Data' category (see Table 1). The contents are also part of the DMP presented by Digital Curation Centre, highlight in the next example.

After summarising the categories of a DMP, other examples are provided next.

Example of a DMP Checklist. DCC [25] synthesises through a checklist that presents the main questions or subjects that researchers may want to cover when writing a DMP (see Fig. 4).

4.2 Recommendations for Developing a Suitable Data Management Plan

Michener [1] recommends ten points for constructing an effective DMP, as summarised in Fig. 5.

Table 1. Categories and their contents for a DMP (based on ref. [22]).

Categories	Brief description
Administrative data	Document basic information in order to identify the plan (e.g. project title and summary), researcher details, applied policies, other procedures, etc.)
Data collection	Describe data collection and procedures to do so (e.g. existing data that can be reused, standards/methodologies to be applied, adopted quality assurance, etc.)
Documentation and metadata	Taken into account what information is needed for the data to be read and interpreted in the future (e.g. estimate how much time and effort are needed, establish documents and metadata that goes together with the data, metadata standards to be adopted, including reasons to do so, etc.)
Ethical and legal compliance	Consider any ethical, legal issues, or constraints that may be a concern for data sharing (e.g. data consent for sharing and preserving, protection for participants identification, data sharing postponement/restriction (e.g. related to patents, licensing data for reuse, etc.)
Storage and backup	Taken into account location of data storage and any implications this may have for backup, access, and security (e.g. sufficiency of storage, responsibility for backup/recovery, risk management, ensure secure access to people involved, etc.)
Selection and preservation	Determine which data are of long-term and should be preserved, including decision of best ways to preserve data (e.g. by depositing in repositories, data that must be retained/destroyed for contractual, legal, or regulatory purposes, expected research data uses, conservation and potential sharing, time and effort costing needed to prepare the data for preservation/sharing)
Data sharing	Consider which data that are to be shared and means to do so by taking into account that methods are dependent on various factors such as type, size, complexity, and sensitivity, etc. Also taken into account how people might acknowledge the reuse of data (e.g. citations) to gain impact (with whom to share the data and under what conditions, decision on making the data available, restrictions on data sharing, etc.)
Responsibilities and resources	Assign roles and responsibilities for every one involved in data management. Additionally, any resources needed to deliver the plan should be weighed up (e.g. DMP implementation responsibility, assure revision, split responsibilities among partners, needed resources to deliver the plan, additional specialist expertise/equipment, etc.)

PROJECT IDEA TEMPLATE

Title & acronym of the project	included4DI – Inclusive Practices for Diversity and Integration
Programme	ERASMUS +
Key action	Key Action 2: Cooperation for innovation and the exchange of good practices
Action	Strategic Partnerships
Duration of the project	24 months
Deadline	26/02/2020
Budget	Maximum EU contribution awarded per 1 year of project implementation is 150 000 EUR 250.000
Developing organization	University of Aveiro
Contact person	Marlene Amorim
Erasmus+ priorities	Horizontal: Social Inclusion Specific: Building inclusive higher education systems
Target group/s	Direct: Students, Educators indirect:

SUMMARY/RATIONAL

The development of inclusive development is a foundational element the global agendas in contemporary societies. However, it requires continuous reflection and learning in order to improve our individual and organizational practices in what concerns the relationships and representation of human disabilities within work to achieve transformative, systemic change in work and life contexts.

context and profiles, and build an awareness pack consisting in content and materials, in the form of cases and examples of diversity and disabilty inclusion challenges associated with communication in education and work contexts, building on the collection of evidence from accross the partner countries [O1]; ii) develop methods and tools to assist educators and recruiters in the development of inclusive communcation and content, including guidelines and examples of inclusive communcation, practical checklists to assess the inclusiveness of content of communication materials, to meet a set of selected/relevant profiles and contexts of diversity and disability [O2]; develop an online tool to support and make available the elements resulting from O1 and O2, and to integrate a tool to support the online and interactive assessment of the quality of communication materials and content produced by educators and recruiters [O3]; create learning/training materials for educators, trainers and recruiters to deploy the use of O1, O2 and O3, pilot these materials and make them available to wider audiences [O4]; conduct a short training event/pilot involving participants from partner countries, and engaging representatives from the target audiences, including educators, trainiers and recruiters.

PROJECT OUTPUTS/ INTELLECTUAL OUTPUTS
Please list the outputs to be developed along with a brief description for each one.

IO1: Diversity and disability profiles and content tipology and characterization method

IO2: Methodology and guide to identify diversity and disability gaps in communication and content materials

IO3: Online tool to disseminate diversity and disability integration tools and materials and incorporating online interactive function to assess the quality of communication materials

IO4: Tool kit for training educators, trainers and recruiters.

PROPOSED PARTNERSHIP
List here any organisation that has been already included in the partnership

A/A	ORGANIZATION	COUNTRY	ROLE	MAIN RESPONSIBILITIES
1	Uni Aveiro	Portugal	APPLICANT	IO1

Diversity and disability inclusion is part of a wider ambition for inclusive development that strives for the active participation and representation of all people regardless of age, gender, disability, ethnicity, race, class, religion, sexuality or any other characteristic.

The project include4DI – Inclusive Practices for Diversity and Integration aims to contribute to the development and. adoption of inclusive practices for in education, training and job integration contexts, notably trough the development of knowledge, tools and sharing practices that promote inclusive communication, anf foster the access and integration of individuals in a diverse and representative work and like context.

Estimating an accurate figure of the number of people with some form of disability is a challenging task to start with. Differences exist in the ways countries define disability; the quality and methods of data collection; reliability of sources; and disclosure rates.The Worls Report on Disability is a reference source in the topic and advances figure over a billion people, about 15% of the world's population, who have some form of disability. Disability can be defined as '… an evolving concept that results from the interaction between persons with impairments and attitudinal and environmental barriers that hinders their full and effective participation in society on an equal basis with others' - United Nations Convention on the Rights of Persons with Disabilities – or, in other words, disability can be addressed as the relationship between a person's impairment and their environment. As a human rights issue disability shares many common experiences with other groups that have been traditionally marginalised or excluded.

AIM AND OBJECTIVES

The project include4DI – Inclusive Practices for Diversity and Integration aims to create and disseminate knowledge and tools that can support the development of more inclusive communication prectives in education, qualification and job integration contexts, in order to promote a wider access, integration and representation of diverse individuals in education, work and life contexts, facilitating the elimination of barriers to equal access to opportunities.

In order to achive its goals the project sets up to develop a toolkit and training tools to foster the inclisive communition and integration of individuals with diverse backgrounds and disability characteristics, including: i) develop a framework to identify and map diversity and disability

2	EUROSUCCESS	Cyprus	PARTNER	IO3
3	Dom(Spain)	Spain	PARTNER	IO2
4			PARTNER	
5			PARTNER	
6			PARTNER	

PARTNERS NEEDED

Describe a profile/s of the partner/s you are looking for…

The project is looking for partners from countries/contexts where cultural diversity is prevalent, where education and work contexts are addressing challenges in the integration and communication with diversified backgrounds and individual conditions (e.g. migrants).

Fim do documento ■

Fig. 3. Project idea template example.

Earlier points on DMP raised in Fig. 5 are similar than other sources. However, the author adds more specific details on procedures, tools, etc. as a useful guidance to the development of a DMP. It is worth noting that previous description is an extraction. Further contents can be found in ref. [1].

DCC Checklist	DCC Guidance and questions to consider
Administrative Data	
ID	A pertinent ID as determined by the funder and/or institution.
Funder	State research funder if relevant
Grant Reference Number	Enter grant reference number if applicable [POST-AWARD DMPs ONLY]
Project Name	If applying for funding, state the name exactly as in the grant proposal.
Project Description	**Questions to consider:** - What is the nature of your research project? - What research questions are you addressing? - For what purpose are the data being collected or created? **Guidance:** Briefly summarise the type of study (or studies) to help others understand the purposes for which the data are being collected or created.
PI / Researcher	Name of Principal Investigator(s) or main researcher(s) on the project.
PI / Researcher ID	E.g ORCID http://orcid.org/
Project Data Contact	Name (if different to above), telephone and email contact details
Date of First Version	Date the first version of the DMP was completed
Date of Last Update	Date the DMP was last changed
Related Policies	**Questions to consider:** - Are there any existing procedures that you will base your approach on? - Does your department/group have data management guidelines? - Does your institution have a data protection or security policy that you will follow? - Does your institution have a Research Data Management (RDM) policy? - Does your funder have a Research Data Management policy? - Are there any formal standards that you will adopt? **Guidance:** List any other relevant funder, institutional, departmental or group policies on data management, data sharing and data security. Some of the information you give in the remainder of the DMP will be determined by the content of other policies. If so, point/link to them here.
Data Collection	
What data will you collect or create?	**Questions to consider:** - What type, format and volume of data? - Do your chosen formats and software enable sharing and long-term access to the data? - Are there any existing data that you can reuse? **Guidance:** Give a brief description of the data, including any existing data or third-party sources that will be used, in each case noting its content, type and coverage. Outline and justify your choice of format and consider the implications of data format and data volumes in terms of storage, backup and access.
How will the data be collected or created?	**Questions to Consider:** - What standards or methodologies will you use? - How will you structure and name your folders and files? - How will you handle versioning? - What quality assurance processes will you adopt? **Guidance:** Outline how the data will be collected/created and which community data standards (if any) will be used. Consider how the data will be organised during the project, mentioning

Fig. 4. Example of part of a checklist for a DMP (v.4.0, 2014) [25].

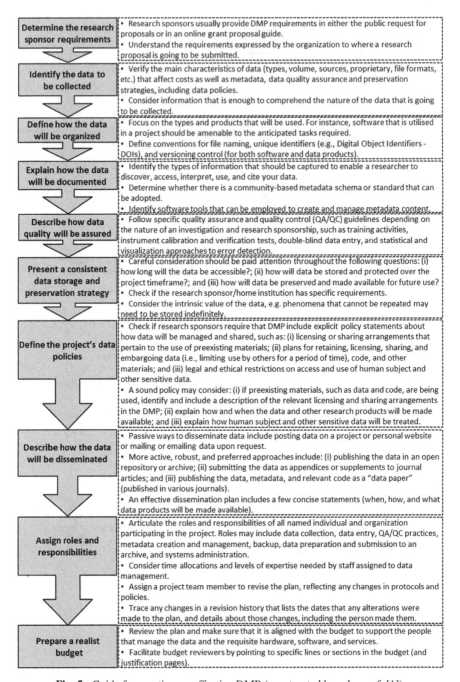

Fig. 5. Guide for creating an effective DMP (constructed based on ref. [1]).

5   Concluding Remarks

Currently, it is undeniable the sharp exponential growth in the amount of data. Thus, it is of paramount importance to ensure that researchers possess the knowledge and to employ best practices with regard to their data during their research as well as after completion. Concerning on all stages of data life cycle is crucial in research, chiefly taking into account the demands from funders.

This work concisely explores the characteristics of the life cycle of data management for research. At this point of this work, the authors expect to start (and expand) this discussion to the research practices in the field of industrial engineering because this is usually omitted, methodologically speaking. As illustrated in the present study, a data management plan is one of the important instruments and should be considered with care when designing and applying it. In this context, it is also relevant to provide a comprehensive support for managing research data generation, preservation, and sharing for reproducibility as well as for further applications. This may include instructions on best practices in data management to graduate students, which is relatively rare in developing countries like Brazil, for instance. Future work may consider providing a structure and guidance for research students in the field of industrial engineering as a valuable avenue to explore.

References

1. Michener, W.F.: Ten simple rules for creating a good data management plan. PLoS Comput. Biol. **11**(10), e1004525 (2015). https://doi.org/10.1371/journal.pcbi.1004525
2. Davis, H.M., Cross, W.M.: Using a data management plan review service as a training ground for librarians. J. Librariansh. Sch. Commun. **3**(2), eP1243 (2015). https://doi.org/10.7710/2162-3309.1243
3. Antell, K., Foote, J.B., Turner, J., Shults, B.: Dealing with data: science librarians' participation in data management at an association of research libraries institutions. Coll. Res. Libr. **75**(4), 557–574 (2014). https://doi.org/10.5860/crl.75.4.557
4. Surkis, A., Read, K.: Research data management. J. Med. Libr. Assoc. JMLA **103**(3), 154–156 (2015). https://doi.org/10.3163/1536-5050.103.3.011
5. Cox, A.M., Tam, W.W.T.: A critical analysis of lifecycle models of the research process and research data management. Aslib J. Inf. Manage. **70**(2), 142–157 (2018). https://doi.org/10.1108/AJIM-11-2017-0251
6. Vieira, R., Ferreira, F., Barateiro, J., Borbinha, J.: Data management with risk management in engineering and science projects. New Rev. Inf. Netw. **19**(2), 49–66 (2014). https://doi.org/10.1080/13614576.2014.918519
7. Anonymous: Nature editorial - making plans. Nature **555**(7696), 286 (2018)
8. Mannheimer, S.: Toward a better data management plan: the impact of DMPs on grant funded research practices. J. eSci. Librariansh. **7**(3), 5 (2018). https://doi.org/10.7191/jeslib.2018.1155
9. Bellgard, M.I.: ERDMAS: an exemplar-driven institutional research data management and analysis strategy. Int. J. Inf. Manage. **50**, 337–340 (2020). https://doi.org/10.1016/j.ijinfomgt.2019.08.009
10. Wright, A.: Electronic resources for developing data management skills and data management plans. J. Electron. Resour. Med. Libr. **13**(1), 43–48 (2016). https://doi.org/10.1080/15424065.2016.1146640

11. Holles, J.H., Schmidt, M.L.: Graduate research data management course content: teaching the Data Management Plan (DMP). In: 2018 ASEE Annual Conference and Exposition (2018)
12. Reilly, M., Dryden, A.R.: Building an online data management plan tool. J. Librariansh. Sch. Commun. 1(3), eP1066 (2013). https://doi.org/10.7710/2162-3309.1066
13. Van Loon, J.E., Akers, K.G., Hudson, C., Sarkozy, A.: Quality evaluation of data management plans at a research university. IFLA J. 43(1), 98–104 (2017). https://doi.org/10.1177/034003 5216682041
14. European Commission – European Union. https://ec.europa.eu/research/participants/ docs/h2020-funding-guide/cross-cutting-issues/open-access-data-management/data-manage ment_en.htm. Accessed 11 Jan 2020
15. Willaert, T., Cottyn, J., Kenens, U., Vandendriessche, T., Verbeke, D., Wyns, R.: Research data management and the evolutions of scholarship: policy, infrastructure and data literacy at KU Leuven. LIBER Q. 29, 1–19 (2019). https://doi.org/10.18352/lq.20272
16. Elsevier. https://researcheracademy.elsevier.com/research-preparation/research-data-manage ment. Accessed 11 Jan 2020
17. RDMLA – RDMLA. https://rdmla.github.io/about/. Accessed 11 Jan 2020
18. Wilkinson, M.D., et al.: The FAIR guiding principles for scientific data management and stewardship. Sci. Data 3, 160018 (2016). https://doi.org/10.1038/sdata.2016.18
19. Booth, A., Sutton, A., Papaioannou, D.: Systematic approaches to a successful literature review. Sage Publications Ltd., Thousand Oaks (2012)
20. European Commission. https://ec.europa.eu/research/participants/docs/h2020-funding- guide/cross-cutting-issues/open-access-data-management/data-management_en.htm. Accessed 9 Jan 2020
21. Digital Curation Centre –DCC. http://www.dcc.ac.uk/. Accessed 8 Jan 2020
22. Digital Curation Centre –DCC. http://www.dcc.ac.uk/sites/default/files/documents/resource/ DMP/DMP-checklist-flyer.pdf. Accessed 6 Jan 2020
23. University of Aveiro. https://www.ua.pt/pt/investigacao. Accessed 9 Jan 2020
24. European Commission. https://ec.europa.eu/programmes/erasmus-plus/about_en. Accessed 9 Jan 2020
25. Digital Curation Centre – DCC. http://www.dcc.ac.uk/sites/default/files/documents/resource/ DMP/DMP_Checklist_2013.pdf. Accessed 6 Jan 2020

A Blockchain Approach to Social Responsibility

Augusto R. C. Bedin[1], Wander Queiroz[1], Miriam Capretz[1(✉)], and Syed Mir[2]

[1] Western University, London, ON N6A 5B9, Canada
{arodr3,wqueiroz,mcapretz}@uwo.ca
[2] London Hydro, London, ON N6A 4H6, Canada
mirs@londonhydro.com

Abstract. As blockchain technology matures, more sophisticated solutions arise regarding complex problems. Blockchain continues to spread towards various niches such as government, IoT, energy, and environmental industries. One often overlooked opportunity for blockchain is the social responsibility sector. Presented in this paper is a permissioned blockchain model that enables enterprises to come together and cooperate to optimize their environmental and societal impacts. This is made possible through a private or permissioned blockchain. Permissioned blockchains are blockchain networks where all the participants are known and trust relationships among them can be fostered more smoothly. An example of what a permissioned blockchain would look like is described in this paper as well as its implementation, achieved using Hyperledger Fabric, which is a business-oriented blockchain framework. This study touches on the benefits available for companies that are willing to engage in socially responsible causes through blockchain. It states in what ways a permissioned blockchain can bring together businesses on common ground to increase their reach and provide better customer service. Finally, a use case is provided to bring to life a real-world situation where blockchain use improves service quality for all the parties involved, both the companies and their customers.

Keywords: Blockchain · Social responsibility · Hyperledger Fabric · Permissioned blockchain

1 Introduction

Blockchain is still riding on the hype generated by the cryptocurrency fever after it revolutionized the way that money can be traded, created, and earned. Consequently, blockchain is commonly considered to be a finance-oriented technology because of its inherent traits of security and integrity in handling data. Its prominent and highly regarded success cases Bitcoin and Ethereum are both supporting proofs that blockchain can yield impressive outcomes when and if properly used. The big cryptocurrency blowout led to blockchain technology

© ICST Institute for Computer Sciences, Social Informatics and Telecommunications Engineering 2020
Published by Springer Nature Switzerland AG 2020. All Rights Reserved
R. Mugnaini (Ed.): DIONE 2020, LNICST 319, pp. 15–29, 2020.
https://doi.org/10.1007/978-3-030-50072-6_2

being introduced into various sectors, evolving and shaping its usage to fit the required purposes. Miraz and Ali [19] point out that blockchain can very well provide in various scenarios other than financial related ones - any circumstance that would require a high level of trust among parties or even a third party intermediate to validate the interaction can benefit from blockchain's trusted environment. Owing to this, blockchain found its way into energy [4], government [23], IoT [9], and even medical [28] industries demonstrating its flexibility in numerous scenarios.

One of the relatively unexplored paths that blockchain can take is that of social responsibility. It is in the best interest of companies to seek a balance between economic prosperity and societal issues to perform efficiently and effectively. Social responsibility is, as the International Organization for Standardization (ISO) states, the responsibility of an organization for the impacts of its decisions and activities on society and the environment through transparent and ethical behaviour [13]. Social responsibility [20] can be referred to as the ability of a company to focus on issues beyond profitability, extending further than its economic frontiers to touch on ethical, legal, and philanthropic matters. Profit-focused companies are known to dread social responsibilities on the account that it may lead to additional productions costs. Further, Morh *et al.* [20] state that a reliable percentage of customers (49%) are inclined to select a more socially responsible company over its less socially aware competitors, encouraging companies to do so as well as enforcing the relevance of this presented work.

As far as blockchain technology goes, applications with that purpose are still scarce, under development, and far from being thoroughly examined. As a distributed ledger technology (DLT), blockchain technology presents itself as a means to tighten bonds between organizations and its customers by providing more data integrity, security, and transparency for its services. Additionally, permissioned blockchains have a unique knack for handling business activities very well. With a more business-centered attitude, permissioned blockchains enable organizations to structure access levels for its members through valid credentials. When this is achieved, an organization's hierarchical system can be installed on the blockchain to do business through a safeguarded medium. Such a feature, which is only available on permissioned blockchains, can be a useful tool for companies to implement social responsibility. Naturally, it is reasonable to contemplate the likelihood of building such an application without adopting blockchain technologies. Although possible, the technical complexity of it quickly reveals itself a barrier. Blockchain grants, upon majority approval, newcomers a swift integration into the network and so to the data available in it. Security and integrity technicalities are not a matter of contention when dealing with blockchain. Nakamoto's first paper on peer-to-peer network [22], which later became known as blockchain, cites how little effort is required not only to set up a blockchain network but also how easily nodes can join or leave the network at will, accepting that the longest chain of blocks in it as proof of everything that happened when they were absent.

Today's business models are based on catering to potential customers with deals for them so that they can pick the most suitable and highest cost-benefit services they require. This makes sense only from a supply and demand point of view. However, this results in customers scattering their personal information around to various companies, creating a large number of failure points where information can be accessed or viewed by unwanted eyes. One way to minimize this issue would be to gather together, on common ground, companies that operate on the same sector or market into a permissioned blockchain, along with the information of their shared customers. In this manner, customers' information would be available to the organizations through a secure, trusted medium.

Furthermore, businesses might have services that they require from other companies. This would then be reflected in shared customers for these companies. Indirectly related businesses could benefit from tighter interaction. Because in traditional settings the customer must interact with possibly multiple companies to obtain the full service that he or she needs, companies that rely on other companies to perform specific tasks could have a smoother client transition if they were aware of what was happening on each end. As an example, if company A would like company B to perform a service, assuming that both are part of a permissioned blockchain, company A could communicate directly with B, provided that clients' consent was provided, without having the customer acting as an intermediary. This responsibility could be taken out of customers' hands and passed on to the organizations themselves to deal with in favour of improved customer satisfaction. As privacy matters are not to be taken lightly in any circumstance, it is imperative for clients to be aware of where, how and why their information are getting to the companies involved. Avoiding compliance to that would void blockchain's transparencies edge.

This paper sheds light on how blockchain can become a strong ally for wise and socially responsible companies. It specifies how permissioned blockchain networks can be tailored to tackle societal issues that stretch beyond an organization's duty in a way from which they can very well benefit. On this relatively uncharted trail, this discussion elicits how permissioned blockchain, along with its features, can tighten the interaction between common-ground business parties and their shared customers for their benefit and ultimately the benefit of the whole society. This paper's contributions support building a permissioned blockchain environment where organizations can work alongside other organizations to uphold a closer relationship through a candid regime established through smart contracts, direct channels of communications, and transparency due to the nature of distributed ledgers. Its essence is to define how clients can more easily get one organization's services to another as seamlessly as possible.

It is particularly worth mentioning that this paper is limited to the business scenario that it addresses. Its actual contribution acts on strengthening bonds among companies and would be impaired or even ineffective by the lack of business-to-business relationships.

The remainder of this paper is organized as follows: Sect. 2 addresses blockchain background and the framework that is used to achieve the results, followed by related work in Sect. 3. The specifics of blockchain design and implementation are outlined in Sect. 4. Section 5 contains an example of a use case applying the method described in Sect. 4 and finally, some thoughts and remarks to conclude the paper.

2   Blockchain Technology

The information available in this section formulates the context in which this study is situated. Terminologies and necessary complementary information on blockchain are reviewed to avoid misinterpretation. Additionally, pertinent blockchain and social responsibility studies are also discussed here to support and justify this paper's endeavors.

2.1   Background

Blockchain technology consists of recording the exchange of information within a network where every user holds a ledger with the records of every movement of every piece of information [12]. This information is exchanged in the form of authorized transactions validated by the network users. An arbitrary number of transactions can then be packed along with an identifier (by default, a hash), establishing a block of information. The blocks are created with the sequential identifier of the previous block and put into a chain of blocks. Once the block is placed into the chain, its inner data become immutable and can no longer be tampered with. Because every piece of information inside the blocks has been previously approved by the majority of users in the network, the data are regarded as legitimate and trustworthy. This process is depicted in Fig. 1.

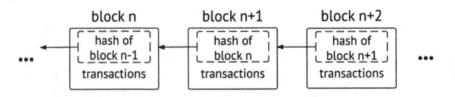

Fig. 1. Blockchain: each block contains an identifier hash pointing to the previous block and a package of transactions that have taken place in the network [6].

Figure 1 illustrates the primary blockchain *modus operandi*. The technology is flexible enough to be tailored to handle the desired information that is relevant to its users. In other words, the ledgers on every user's node will contain information that is pertinent to the whole body of users.

– **Permissionless Blockchain** - The scenario depicted above holds for permissionless, public blockchains. Public blockchain networks allow virtually anyone to join, participate, and access the information within them. Granted, this type of blockchain is presumably large in scale, and therefore to control its anonymous and untrusted nodes, a consensus must be reached either through the approval of a majority of users or through a Proof of Work (PoW) [5]. The PoW is evidence that a user has spent enough computational effort to validate its transaction.

– **Permissioned Blockchain** - Private, permissioned blockchains are generally smaller in scale. In permissioned blockchains, to join, one must hold a valid certificate issued by an accredited entity in the network. This normally means that an administrator node must assign an identity for newcomers so that they are no longer anonymous, and other users/nodes can recognize their actions on the blockchain. Permissioned blockchains act as a semi-trusted setting [5], and because of that, consensus can be reached differently than on permissionless, public blockchains. Consensus on permissioned settings is not as costly to achieve as on permissionless blockchains since nodes are identifiable and trusted within that network. No proof of work is required for the nodes to invoke transactions. The only necessary verification is the validity of the data within a node's transaction.

– **Smart Contracts** - Smart contracts are, in a blockchain frame of reference, code scripts stored within the blockchain [6]. They are triggered autonomously and fundamentally by transactions in a structured way throughout all the nodes in the network, conforming to the data used by the triggering transaction. Figure 2 illustrates how a smart contract execution process works. When invoking a transaction that triggers a smart contract its code will run in every node involved in the transaction, executing the business logic that it was designated for. Smart contracts enable a more fluid work flow by prompting computational procedures on demand, responding involved parties under a structured policy. By doing so, they allow a reduced number of trusted intermediaries to be involved in parties' transactions while also minimizing accidental and erroneous transactions.

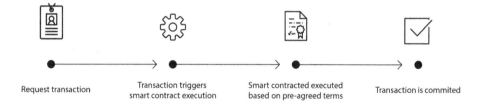

Request transaction   Transaction triggers   Smart contracted executed   Transaction is commited
                           smart contract execution   based on pre-agreed terms

Fig. 2. How smart contracts are triggered and executed on a valid transaction.

2.2 Hyperledger Fabric Framework

The results obtained and described in this paper were achieved using Hyperledger Fabric [11]. Hyperledger Fabric is an open-source permissioned blockchain framework that has been promoted by the Linux Foundation [3] since 2015, and that is fine-tuned with business needs in mind. Hyperledger Fabric, when compared to Ethereum [2] and Corda [1], which are the most known and used blockchains for private purposes, stands out for its scalability [27], flexibility, and customization properties [26]. It is structured to enable high levels of transparency among its users, reinforcing the trust relationship among them. Moreover, because each enterprise on the network holds its ledger, information can be quickly shared and accessed by every participant.

On Fabric, each participant must hold an official identity issued by a trusted Membership Service Provider (MSP) within the network. Having an identity means that the user holds a valid certificate to operate within the blockchain. Having an MSP from an organization in the network to endorse its employees is what differentiates Hyperledger Fabric from other blockchain frameworks. For instance, if organization X has an MSP to issue identities on the network, X's employees can be issued a valid certificate to operate within the network. This enables companies to have a reliable communication channel as well as to log their actions in a transparent and secure environment. When an organization joins a Fabric network, it is called a participant [11]. From now on, the term *participant* will be used to describe an authoritative entity within a Hyperledger Fabric blockchain.

Besides, Fabric allows private and public communication routes to be instituted for participants to converse without sacrificing the proper ordering of network activities, avoiding disparities in a transaction and block timelines. Nodes can be categorized as peer nodes and orderer nodes. Peers generally hold ledgers of information and can request transactions. Peers are the ones who must hold a valid MSP certificate to perform actions on the network. Orderers are nodes that are in charge of adequately ordering transactions in a timely manner, as well as ordering the generated blocks that hold them. Because they cannot request any transactions, they do not require identities to be on the network.

Note that in Fabric smart contracts are called *chaincodes* [11]. From this point on, smart contracts will be referred to as chaincode.

The block generation process on Fabric is slightly different from other frameworks. These differences are highlighted in its block generation process, which is composed of three phases: proposal, ordering and packing, and validation and commit, as depicted in Fig. 3.

1. In the first phase (proposal), an application or participant requests authorization to invoke chaincode from the respective endorsing peers in the network to check whether they agree with the outcome of this chaincode. If they do, they send a response back with their digital signature approving the request. If a majority of the endorsing peers do not sanction it, the request gets rejected.
2. Ordering and packing rely on packaging the approved transactions into blocks, a task that is performed by the ordering node. Ordering nodes do not execute any chaincode and are specially designed to order the blocks properly in the blockchain. The transactions are not always ordered in the same sequence that they are received by the ordering node because multiple ordering nodes can receive transactions simultaneously. However, it is worth mentioning that

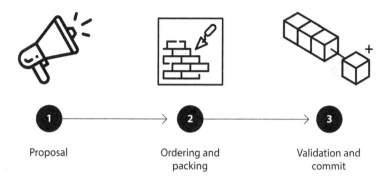

Fig. 3. Simplified visualization of the process described in Subsect. 2.2. **1** - The proposal phase notifies the concerning peers of its intent. **2** - The ordering and packing phase is responsible for correctly building the block to be added to the chain. **3** - The validation and commit phase checks if the block's information checks out and if so, adds it to the blockchain.

the ordering nodes follow a systematic procedure to pack the transactions into blocks. Once the packing of a transaction into a block is done, it is definitive. In Fabric, its position in the chain is irrefutably assured at this point, which is not always true for other blockchain frameworks because some other frameworks may require additional transaction validation after a transaction has been inserted into a block. Fabric assumes that the transactions within a block are absolute because they were previously endorsed by the majority of the network's peers.

3. For validation and commit (the final phase), the blocks are disseminated to other connected peers, and each of them processes the received block to maintain consistency. Once a peer secures a block, it then verifies all the transactions within it before committing them to its ledger to ensure that none of them has been invalidated midway. This may occur because the transaction might have already been entered into the ledger, and therefore adding it again would result in an inconsistency fault. Once everything has been checked, the ledger is updated with a new block of transactions, keeping invalid transactions for audit purposes. The addition of a new block in the blockchain triggers an event to summarize the newly added information to all the peers. Events are also an indicator that block addition processes have been performed. The blocks are organized in the same way as depicted in Fig. 1.

Ultimately, the information stored in blocks and moved around through transactions within Fabric is called an *asset*. An asset is an information that the network users perceive as valuable. It is the main focus of the network to exchange that information and to make it available on all the ledgers for users to access, as illustrated in Fig. 4.

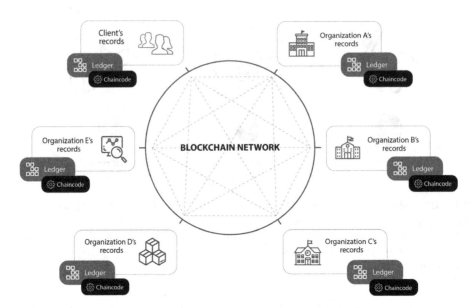

Fig. 4. Organizations and customers join a blockchain to have access to a shared ledger of information. Each participant is a valid peer that holds a copy of the shared ledger. Also, each participant can execute its designated chaincode to perform actions on the network.

Another concern addressed in Hyperledger Fabric is the time complexity of its performance. Blockchain technology inherently sacrifices its read and transaction times for security. Latency on transaction and read times can often take seconds or even minutes on traditional permissionless blockchains. To avoid that, Hyperledger Fabric stores its world state (the most recent state of every node) on a quickly accessible nested database on every node. This cuts the reading time significantly over the fact that searching for information within the ledgers does not require a linear search down the chain of blocks. Of course, many things should be taken into consideration when evaluating a blockchain operation's time complexity such as the number of nodes, the geographical distribution of nodes, type of data stored, workload, among others described in the Hyperledger Performance and Scale Working Group's whitepaper [24]. Overall, time complexity varies from use case to use case, and an ideal equation has not been determined yet, but a suitable suggestion is to use $Product = TransactionThroughput * log(NetworkSize)$.

3   Related Work

Blockchain for enterprises is significantly on the rise. This is confirmed by the quantity of papers on the topic published in recent years as different applications using blockchain are revealed. Hebert and Di Cerbo [10] outline a methodology for businesses to better reap the benefits of using blockchain in the private sector. It expresses the concerns and advantages of using blockchain while drawing attention to how to mold software architectures to achieve maximum business blockchain efficiency. Various markets, financial and non-financial, have become aware of blockchain benefits, and its implementation has spread to seize new-found opportunities [8].

Addressing healthcare, Liang *et al.* [17] highlighted how permissioned blockchains have presented themselves as a strong ally in dealing with healthcare issues, given their enhanced data security and integrity. Due to the sensitivity and privacy requirements of health data, blockchain has become a powerful tool for safely collaborating and sharing data in this category.

In terms of financial and business data, Chua *et al.* [7] have presented a solid example of how permissioned blockchains, specifically Hyperledger Fabric, can serve as a foundation for business activities. It explains how a group of closely collaborative enterprises can gain quick access to data on a distributed ledger network such as a blockchain. In addition, the study stresses how permissioned blockchains provide access control over their data to enable precise interoperability among the network's participants.

Expanding on permissioned blockchain applications, Kirillov *et al.* [14] make a case for Hyperledger Fabric usage for government purposes, in particular for e-voting. They propose an e-voting blockchain model to increase trust among network participants. Again, the handling of sensitive data such as government information demands a secure and reliable environment, one that blockchain can easily provide.

The examples portrayed above are just a few examples of environments where a permissioned blockchain could be helpful. As stated previously, social responsibility-focussed blockchain applications are still scarce. Use of a blockchain in support of a social cause was examined in Liu *et al.* [18], where a blockchain was used to store carbon footprint emissions in Taiwan. The immutability and irreversibility of blockchain data provide a reliable window to check a company's carbon emissions, encouraging more compliant behavior towards environmental regulations and hence towards society overall.

As blockchain applications mature, more complex scenarios can be tackled. On such trend, Li [16] focuses on a philanthropic logistics platform that takes advantage of blockchain's high-profile transparency and credibility. Using Ethereum, the study focused on implementing an application geared towards social welfare maximization, a pertinent example of a social responsibility-targeted blockchain.

Lastly, blockchain geared towards social responsibility is continuously increasing its popularity given social awareness growth in society as general. Mukkamala *et al.*'s study [21] outlines several opportunities and applications for blockchain to enhance social business models in pursuit of maximizing its societal impacts as well as their profits. United Nations Research Institute for Social Development's (UNRISD) working paper [25] highlights blockchain's potential for creating cooperation at scale with cooperative structures for financial inclusion. Kouhizadeh and Sarkis study [15] discuss how blockchain can help supply chains to be more green, reducing waste and thus being more socially aware by providing insights and use cases on the subject.

4 Social Responsibility

Although an organization can strive for social responsibility in many ways, this section discusses how this can be accomplished through permissioned blockchain networks. A use case is also described to clarify how this can be implemented in a real-world setting.

4.1 Design

With the framework that Hyperledger Fabric provides, it is possible to come up with a blockchain design that can reap the advantages of a permissioned blockchain to achieve social responsibility objectives.

Fabric enables straightforward translation of this scenario into a blockchain. In a simplistic approach, companies with that goal in mind can create and join a permissioned blockchain incorporating an MSP within it, turning them into participants. From there, their main asset on the network would be their standard customer information and what kind of service they can provide or require from one another that jointly affects these customers. Transactions and chaincode can be implemented to automate workflow and minimize delays previously experienced when using less efficient communication routes.

Permissioned blockchain networks, and especially Fabric, are conveniently extensible and can accommodate numerous nodes with little effort. Integrating new companies is not only possible but highly encouraged because businesses joining the network can make themselves noticed by existing customers, who are prone to trust newcomers because access to the blockchain is controlled.

It should be pointed out that cooperating with other businesses is not one of the responsibilities of companies. Essentially, the goal is to set up an environment where companies can work actively together for the benefit of their customers. Businesses have the opportunity to help each other grow and profit mutually while increasing customer satisfaction.

Figure 5 presents an example of how a network arrangement can facilitate communication between organizations. Organization A, Organization B, and customers[1] are inserted into a permissioned blockchain network. A customer requests a service from Organization A. Before the service can be delivered, a service from Organization B is required. In traditional settings, Organization A would advise the client to obtain the service from Organization B first and then return to obtain the initially requested service from Organization A, as illustrated in Fig. 5.a. This situation can drastically change when a permissioned

---

[1] Each customer is represented in the blockchain by a single node, meaning that every customer's action in the blockchain network is validated through a single certificate held by the customer node.

blockchain network is used. Within a blockchain, Organization A can promptly communicate with Organization B to obtain the prerequisite service without having the client as an intermediary. The client can then receive the service that he/she initially requested without the hassle of interacting with both companies back and forth, as shown in Fig. 5.b.

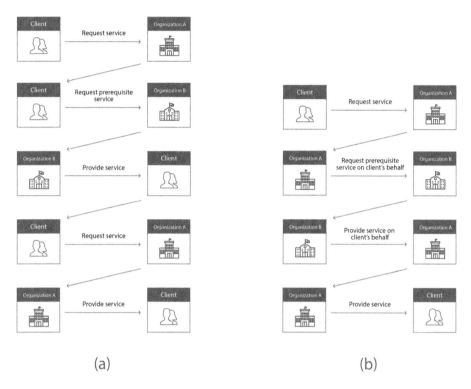

(a)   (b)

Fig. 5. Comparison of service delivery in a traditional setting (a) and through a permissioned blockchain (b).

4.2   Use Case

Section 4.1 described how a permissioned blockchain can bridge the gap between companies and enable them to interact, putting the responsibility of providing a complete service in their hands instead of having the customer chase after it. In this section, a real business situation is depicted to clarify the merits of a permissioned blockchain over traditional approaches.

For the sake of context, assume now that Organization A is a town's local energy provider and that the customers are households that benefit from the provider's service. For numerous reasons, there might be a time where a household is unable to pay its electricity bills, forcing Organization A to perform a service interruption. Between the time of late bill detection and the actual disruption, customers have a grace period to look for assistance programs that can support them. Therefore, it becomes the customer's responsibility to look for help to avoid being disconnected because energy providers and assistance programs work separately. Such a scenario not only puts the customer in danger of being disconnected but also the energy provider in danger of not receiving any payment and the assistance programs in risk of being considered ineffective, turning this into a no-win situation.

In a permissioned blockchain network composed of a local energy provider (Organization A), its customers, and finally an assistance program (Organization B), helping customers in need morphs into a much simpler process. With direct routes of communication, the local energy provider can inform assistance programs directly which customers require aid. Given the distributed ledger nature of blockchains, assistance programs within the network may have access to the local energy provider's list of clients, effortlessly identifying which customers need help. Permissioned blockchains can incorporate organizations smoothly, making it easy for multiple assistance programs to take part in the network and cooperate with the local energy provider and its customers.

The stronger alliance built among entities within the blockchain network and the availability of information are already compelling reasons to use blockchain in situations like that just described. On top of that, using chaincode to automate the matching of a customer in need with the best-fitting eligible assistance program for their case makes the use of permissioned blockchains more persuasive. Once a customer's status shifts to being behind on their bills, chaincode can inform him/her of assistance programs for which they are eligible, enabling them to pick the most suitable one for them. Furthermore, aside from blockchain helping clients better avoid disconnection, it can also produce a promising environment for assistance programs to achieve a broader radius of effectiveness, as well as enhancing the local energy provider's payment collection efficiency. In summary, the full outcome can bring about a win-win situation for all parties involved.

The premise of this use case is to fit the application in a real-world situation with feedback from the company, which would implement it as well as reap its benefits, consolidating its validity as a socially responsible use for blockchain.

5   Conclusions

Blockchain is a multifaceted tool that can be adapted to fit a great variety of purposes. Although most blockchain applications emphasize cryptocurrency and related financial technology, a plethora of other sectors remain fairly untouched by it. There is no reason to believe that organizations that work closely together would not benefit from operating within a permissioned blockchain setting provided that the transparency and security traits of blockchain can profoundly and positively impact the trust relationship between them.

This study has focussed on demonstrating how companies can build social responsibility by taking on, through blockchain, some of their customers' burden of dealing with several companies at a time to receive a single service. By identifying the needs of their customers, companies can limit their downtime by directly reaching the company that provides the service on which the customer depends, instead of having clients as intermediaries. This feature is an extraordinary service that goes beyond the reach of a company's responsibility, but it is also one that can be quite rewarding in terms of both financial profit and reputation.

Also presented here were a structured approach and a real-world scenario illustrating the use of a permissioned blockchain to seize an opportunity to help customers stay up to date with their electricity bills and avoid disconnections. It highlights how working together with recognizable parties is paramount to provide better services to those in need. However, businesses are often reluctant to cooperate because disclosing operational information can be seen as a threat rather than an opportunity for growth. This paper aims to foster the idea of using blockchain as a means to securely unveil information with other companies to mature and progress together. The benefits described here are expected to lead to a more socially wise way of doing business.

Because a permissioned blockchain can centralize customer information, it can become a single means of access to these data. By turning blockchain into an inviting and promising environment for businesses, in terms of future work, it is possible to consolidate a consortium of companies to serve its customers better. Elaborating on this with a real-world scenario, it is possible to gather together companies that provide basic needs such as energy, gas, water, telecommunications, and TV services into a permissioned blockchain to offer their services to households more swiftly (assuming, of course, that the households' information is stored in the blockchain as customer data). On top of that, assistance programs relating to the most essential services such as electricity, water, and gas can be added to the consortium to guarantee service delivery to households in need of aid. Taking advantage of Hyperledger Fabric's ability to enable exclusive routes of communication between the participants in a blockchain, an organized conversation among assistance providers and service providers can be achieved without disrupting other participants. This feature enables a blockchain to manage an increasing number of participants in the network without creating interference.

Given all the customer's data transfer and exchange, privacy, and legal implications, concerns do arise. This scenario is based entirely on customers' consent, willing to allow their information to travel among service providers for their benefit and interest. Indeed, further legal ramifications must be inspected upon the actual implementation of the blockchain network.

With this in mind, one further step in this study would be to explore the possibilities of bringing together multiple business partners into a single blockchain so that customers can make use of a single service request platform. The information that each partner would require from its customers would be elicited to design and develop the appropriate smart contracts that address their needs.

Acknowledgement. This research has been partially supported by an OCE VIP II at Western University (VIP II - 31101). The authors would like to thank London Hydro for supplying industry knowledge used in this study.

References

1. Corda. https://www.corda.net/. Accessed 13 Jan 2019
2. Ethereum. https://www.ethereum.org/. Accessed 13 Jan 2019
3. Linux Foundation. https://www.linuxfoundation.org/. Accessed 13 Jan 2019
4. Andoni, M., et al.: Blockchain technology in the energy sector: a systematic review of challenges and opportunities. Renew. Sustain. Energy Rev. **100**, 143–174 (2019)
5. Baliga, A.: Understanding blockchain consensus models (2017)
6. Christidis, K., Devetsikiotis, M.: Blockchains and smart contracts for the Internet of Things. IEEE Access **4**, 2292–2303 (2016)
7. Chua, P.H.T., Li, Y., He, W.: Adopting hyperledger fabric blockchain for EPC-global network. In: 2019 IEEE International Conference on RFID (RFID), pp. 1–8. IEEE (2019)
8. Crosby, M., Pattanayak, P., Verma, S., Kalyanaraman, V., et al.: Blockchain technology: beyond bitcoin. Appl. Innov. **2**(6–10), 71 (2016)
9. Dorri, A., Kanhere, S.S., Jurdak, R., Gauravaram, P.: Blockchain for IoT security and privacy: the case study of a smart home. In: 2017 IEEE International Conference on Pervasive Computing and Communications Workshops (PerCom Workshops), pp. 618–623. IEEE (2017)
10. Hebert, C., Di Cerbo, F.: Secure blockchain in the enterprise: a methodology. Perv. Mob. Comput. **59**, 101038 (2019)
11. Hyperledger: Hyperledger fabric documentation, release 1.4 (2019). https://hyperledger-fabric.readthedocs.io/en/release-1.4/
12. Iansiti, M., Lakhani, K.R.: The truth about blockchain. Harvard Bus. Rev. **95**(1), 118–127 (2017)
13. Guidance on social responsibility. Standard, International Organization for Standardization, 11 March 2010
14. Kirillov, D., Korkhov, V., Petrunin, V., Makarov, M., Khamitov, I.M., Dostov, V.: Implementation of an e-voting scheme using hyperledger fabric permissioned blockchain. In: Misra, S., et al. (eds.) ICCSA 2019. LNCS, vol. 11620, pp. 509–521. Springer, Cham (2019). https://doi.org/10.1007/978-3-030-24296-1_40
15. Kouhizadeh, M., Sarkis, J.: Blockchain practices, potentials, and perspectives in greening supply chains. Sustainability **10**(10), 3652 (2018)
16. Li, J.: Public philanthropy logistics platform based on blockchain technology for social welfare maximization. In: 2018 8th International Conference on Logistics, Informatics and Service Sciences (LISS). IEEE (2018)
17. Liang, X., Zhao, J., Shetty, S., Liu, J., Li, D.: Integrating blockchain for data sharing and collaboration in mobile healthcare applications. In: 2017 IEEE 28th Annual International Symposium on Personal, Indoor, and Mobile Radio Communications (PIMRC), pp. 1–5. IEEE (2017)
18. Liu, K.-H., Chang, S.-F., Huang, W.-H., Lu, I.-C.: The framework of the integration of carbon footprint and blockchain: using blockchain as a carbon emission management tool. In: Hu, A.H., Matsumoto, M., Kuo, T.C., Smith, S. (eds.) Technologies and Eco-Innovation Towards Sustainability I, pp. 15–22. Springer, Singapore (2019). https://doi.org/10.1007/978-981-13-1181-9_2
19. Miraz, M.H., Ali, M.: Applications of blockchain technology beyond cryptocurrency. arXiv preprint arXiv:1801.03528 (2018)
20. Mohr, L.A., Webb, D.J., Harris, K.E.: Do consumers expect companies to be socially responsible? The impact of corporate social responsibility on buying behavior. J. Consum. Aff. **35**(1), 45–72 (2001)

21. Mukkamala, R.R., Vatrapu, R., Ray, P.K., Sengupta, G., Halder, S.: Blockchain for social business: principles and applications. IEEE Eng. Manag. Rev. **46**(4), 94–99 (2018)
22. Nakamoto, S.: Bitcoin: a peer-to-peer electronic cash system. Technical report, Manubot (2019)
23. Ølnes, S., Ubacht, J., Janssen, M.: Blockchain in government: benefits and implications of distributed ledger technology for information sharing (2017)
24. Performance, H., Group, S.W.: Hyperledger blockchain performance metrics (2018)
25. Scott, B.: How can cryptocurrency and blockchain technology play a role in building social and solidarity finance? Technical report, UNRISD Working Paper (2016)
26. Valenta, M., Sandner, P.: Comparison of ethereum, hyperledger fabric and corda. (ebook) Frankfurt School, Blockchain Center (2017)
27. Vukolić, M.: Hyperledger fabric: towards scalable blockchain for business. Trust in Digital Life (2016)
28. Xia, Q., Sifah, E.B., Asamoah, K.O., Gao, J., Du, X., Guizani, M.: Medshare: trust-less medical data sharing among cloud service providers via blockchain. IEEE Access **5**, 14757–14767 (2017)

A Method for Clustering and Predicting Stocks Prices by Using Recurrent Neural Networks

Felipe Affonso[1]([✉]), Thiago Magela Rodrigues Dias[1], and Adilson Luiz Pinto[2]

[1] Centro Federal de Educação Tecnológica de Minas Gerais, Belo Horizonte, Brazil
felipe-affonso@hotmail.com, thiagomagela@gmail.com
[2] Universidade Federal de Santa Catarina, Florianópolis, Brazil
adilson.pinto@ufsc.br

Abstract. Predicting the stock market is a widely studied field, either due to the curiosity in finding an explanation for the behavior of financial assets or for financial purposes. Among these studies the best techniques use neural networks as a prediction tool. More specifically, the best networks for this purpose are called recurrent neural networks (RNN) and provide an extra option when dealing with a sequence of values. However, a great part of the studies is intended to predict the result of few stocks, therefore, this work aims to predict the behavior of a large number of stocks. For this, similar stocks were grouped based on their correlation and later the algorithm K-means was applied so that similar groups were clustered. After this process, the Long Short-Term Memory (LSTM) - a type of RNN - was used in order to predict the price of a certain group of assets. Results showed that clustering stocks did not influence the effectiveness of the network and that investors and portfolio managers can use it to simply their daily tasks.

Keywords: Neural networks · Clustering · Stock market · Deep learning

1 Introduction

Predicting stock market movement is a well-known research field. Some studies argue that it is possible to predict stock market movement [7, 26]. Based on that, it is natural to use all available resources in our dispose in order to confirm these hypotheses. The latest discoveries on the artificial intelligence (AI) area shows that algorithms are able to learn by themselves, recognize patterns, find the best features for a model and many others [17]. By using some deep learning (an AI subarea) methods it is possible to understand financial market better than just reading and looking the data by itself.

Deep Neural Networks have been used several times in order to predict the financial market movement. Studies show that this method is obtaining 48%–54% right predictions [10]. Thus, we can certify that this field of research can be improved. Overall, the key advantage of deep learning is the ability for feature abstraction and for detecting highly complex interactions between these features – resulting in state-of-the-art performance across many applications [14]. Recurrent Neural Network (RNN) is a type of neural

R. Mugnaini (Ed.): DIONE 2020, LNICST 319, pp. 30–40, 2020.
https://doi.org/10.1007/978-3-030-50072-6_3

network which offers the best outcomes when dealing with sequential data, for example, stock returns.

However, using neural networks in order to predict the stock market is a complex and expensive task. It is necessary to prepare the data, prepare the network, adjust the parameters, evaluate used metrics and others. Therefore, the main purpose of this study is verifying if it is possible to forecast the movement of a group of similar stocks. If so, investors and portfolio managers will be able to conduct their investments in a much simpler and less expensive environment.

In order to predict the movement of a group of stocks, they must be previously clustered together. There are several recent studies proposing different ways of clustering stocks [3, 12, 21, 22]. Most of them use more than one technique and focus on portfolio management. In this paper, K-Means algorithm will be used once it outperforms other methods in terms of clusters compactness [3, 22].

The remainder of this paper is organized as follows. Section 2 briefly covers some previous studies in this area. Section 3 provides an in-depth discussion of the methodology, explaining which techniques were used in each part of the process. In Sect. 4 the results are presented. Finally, Sect. 5 presents some discussions about most relevant findings and proposes future works.

2 Literature Review

Längkvist et al. (2014) [16] explains the importance of time-series in data mining and how it can be used as a feature to perform better tasks in AI. By adopting unsupervised training methods, it is possible to use time-series as an important feature of the model and consequently achieve better results. The authors also explain that dealing with time-series can be tough, once the data is noisy and contains several dimensions. When techniques to simplify the data are applied, important information can be lost. Different algorithms are shown and explained, each one of them have their own characteristics and purposes. Focusing on the stock market, deep learning strategies can solve most of the problems on dealing with such type of data and provide new approaches that have not been tested yet. The authors also explain how time-series are used in different fields as music recognition, videos, physiological data, and others.

Bini and Mathew (2015) [3] present some data mining techniques used for clustering and data prediction. After explaining the importance of each categories of data mining the paper goal is presented. The work focuses on finding out the best companies in the market using clustering techniques and predicting future stock price using regression techniques. Among the methods used, K-means and Expectation Maximization algorithms performed better when compared to hierarchical and density-based technique. Whereas the prediction algorithm used was the multiple linear regression that is best than simple linear regression.

Kumar and Ravi (2016) [15] describes past studies and surveys on text mining field including several applications of these techniques. They also present the advantages and disadvantages of methods used on these researches. The paper consists in a literature survey on different databases about the text mining subject applied to financial domain. In order to transform a normal data set into a structured format several techniques can be applied. The first step is called pre-process, it plays an important role in text mining tasks. Normally, when the pre-process step is done with excellence, results tend to be superior. Feature selection is the second step, in this part the unnecessary features are removed, later the data set retains only domain-relevant attributes. The authors present several important information about the number of papers and the year of publication of each one, a distribution of methods and approaches, and also a summary containing the most relevant information about the studies.

Cavalcante et al. (2016) [4] proposes a new literature review on recent approaches designed to solve financial market problems. The main purpose of the study is to survey the machine learning methods applied to the financial context published from 2009 to 2016. Besides summarizing the main studies, the authors have also identified the key challenges on this research field. The articles are categorized by their pre-processing, forecasting and text mining methods. Some concepts are detailed trough the article, as the difference between technical and fundamental analysis, and also the machine learning methods: clustering, artificial neural networks, support vector machines, and others. The authors also explain the advantages of using these techniques in financial data and how it can be used in order to achieve better results. One of the key outputs from this paper is the summary presented, where it is possible to verify features as: main goal, application, which inputs were used, the techniques and if they used some trading system or not. It represents an important research once it presents the key publications and the techniques used in several relevant studies in the past years.

Chong et al. (2017) [6] aims to use deep learning techniques for financial market prediction for 38 stocks from the Korean stock market. The authors compare three deep neural networks methods against a standard regressive model and an artificial neural network. The authors utilize the Normalized Mean Squared Error (NMSE), Root Mean Squared Error (RMSE), Mean Absolute Error (MAE) and Mutual Information (MI) as measures to evaluate the prediction performance the methods utilized. As a result, the deep neural network performs better on the training set, but on the test set the autoregressive model performs better. Both of the methods can be joined in order to combine the advantages of each one and eliminate the disadvantages. Also, it is necessary to use more data aiming high quality outcomes.

The research on the finance and neural networks field is motivated because just a few articles on the area can be found. Krauss et al. (2017) [14] writes that "This point can be illustrated with The Journal of Finance, one of the leading academic journals in that field. A search for "neural networks" only produces 17 references whereas the journal has published about two thousand papers during the last thirty years. An even more limited number of papers uses neural network techniques in their empirical studies." According to the paper, this is a unique study because they use three different state-of-the-art machine learning techniques and compare than with a deep learning algorithm. Also, they follow the financial literature and provide a holistic performance evaluation, making it easy for investors to understand the results achieved. And lastly, they focus on a daily investment horizon instead of months and weeks, as normally on the literature. By using daily values, the model can be trained with more data and consequently perform better. The main goal of the paper is to bridge the gap between academic and professional finance, providing a new perspective for both fields equally.

Fischer and Krauss (2018) [8] propose a new method for financial market predictions using deep learning with long short-term memory (LSTM) networks. This work outperforms Krauss et al. (2017) [14] and achieves returns of 0.46% per day. The main goals of the study are: use LTSM networks on the financial area, provide a better understanding about artificial neural networks and summarize the strategies used by the LSTM in choosing between winning and losing stocks.

Nelson et al. (2017) [23] studies the applicability of recurrent neural networks (RNN), in this case the LSTM was used in order to predict stock prices movement for four Brazilian companies. It was possible to observe that this technique outperforms the baselines with few exceptions. It proves that LSTM networks can offer good predictions when compared to other approaches, mainly in terms of accuracy and gains obtained. Even when compared to other finance strategies, like buy and hold, the trained model outperforms traditional methods and offers less risks when observing the maximum losses.

Zhang et al. (2018) [26] created a model called Xuanwu, which utilizes unsupervised pattern recognition methods in order to remove the human factor in the division of training and test samples. Experimental results show that the proposed framework outperforms the best methods in stock movement prediction. On a more detailed explanation, the first fact is that, it is impossible for human to compete with big data algorithms, so, the separation between train set and test set should be made by the algorithm. The training sample differs from other literature models because it works utilizing the probability of a known shape be formed in a fixed duration of time when only the first days of analyze were conducted. Xuanwu also utilizes pattern recognition algorithms to understand the movement of the stock being analyzed, these patterns are called shapes. The authors claim that this framework can be used by small startups that doesn't trade in the stock market so frequently, the analysts could use the method to narrow down they search for good stocks to invest in.

3 Materials and Methods

The data set used in this paper was downloaded from Kaggle[1]. It contains the stock code, date, and values of close, high, low and volume for a given date. The first step is to define the period in which the data must be extracted. Later, it must be analyzed in order to obtain the period that contains the biggest number of stocks traded. Data preparation and handling was entirely conducted in Python 3.5 [24]. The deep learning LSTM network was developed using Keras [5].

In order to achieve the desired results, it was necessary to complete a sequence of steps. First, it is necessary to (i) prepare the data for future utilization, it consists in downloading the data set, extract the data to be used in the prediction and evaluate if the extraction was done successfully. Later, the preprocessed data must be (ii) clustered, it guarantees that it will be possible to predict values from a large data set, once it would take a long time to run a neural network algorithm through thousands of stocks. Therefore, it will be necessary to find similarities between stocks, this will be done

---

[1] https://www.kaggle.com/borismarjanovic/price-volume-data-for-all-us-stocks-etfs.

using K-Means algorithm, which will be applied on a correlation matrix generated by Pearson correlation coefficient metric. Once the data is grouped by its similarities, neural networks algorithms will be implemented to (iii) predict stock market prices. Finally, the results generated in last step will be evaluated.

3.1 Clustering

Mirkin (1996) [20] defines clustering as a mathematical technique designed for revealing classification structures in the data collected in the real-world phenomena. Which means that clustering is the process of creation of clusters of similar objects [3]. K-means clustering aims to partition n observations into k clusters in which each observation belongs to the cluster with the nearest mean [22]. However, the data must be prepared prior to being used by this algorithm.

The stock return was calculated using Eq. 1, where Rt represents the return for a specific period t. Pt is defined as the price of the stock at time t, and P_{t-1} represents the price of that same stock at the instant just before P_t analysis [1]. In this study, the 1-day interval was used.

$$Rt = \ln\left(\frac{p_t}{p_{t-1}}\right) \tag{1}$$

Pearson's correlation coefficient is used to measure how two linear variables are correlated. The result goes from -1 to 1, the signal indicates the direction of the relationship and the value suggests how strong it is. By other side, if the value is zero, there is no linear correlation between the variables [7].

$$C_{i,j} = \frac{\langle R_i R_j \rangle - \langle R_i \rangle \langle R_j \rangle}{\sqrt{\langle R_i^2 - \langle R_i \rangle^2 \rangle \langle R_j^2 - \langle R_j \rangle^2 \rangle}} \tag{2}$$

The math behind Pearson's correlation coefficient is shown in Eq. 2, where (R_i) is the average value of return prices, so the $(R_i R_j)$ can be explained as the average of the sum of the mean log return values of two corresponding stocks. In other words, it is the mean of the mean. The correlation coefficient is measured for all pairs of stocks in both exchanges.

Cluster analysis is an unsupervised method and one of the biggest problems with it is identifying the optimum number of clusters [9, 19]. Therefore, determining the optimum number of clusters for a given data set is extremely important. Some validation method must be used to achieve the appropriate number of clusters. There are several methods that can be applied, but in this paper the "elbow method" will be used. It consists in plotting a graph where the y-axis represents some kind of error metric and the x-axis represents the number of clusters. A marked flattening of the graph suggests that the clusters being combined are very dissimilar, thus the appropriate number of clusters is found at the "elbow" of the graph [13].

After obtaining the correct number of clusters, K-Means algorithm was applied on all of the correlation matrices and the output was a file containing the stock and its correspondent group.

3.2 Neural Networks

A Neural network consists of an interconnected group of artificial neurons, while deep learning, a method of machine learning, has developed several layers on the basis of neural networks [18]. These networks have been used to achieve state-of-the-art results on a number of benchmark data sets and for solving difficult AI tasks [16].

As we are dealing with stock prices, the data set consists of just an array of numbers through time. Recurrent Neural Networks (RNN) are mainly used to predict a data sequence, for example the next word in a text or message. RNNs process an input sequence one element at a time, maintaining in their hidden units a 'state vector' that implicitly contains information about the history of all the past elements of the sequence [17]. The RNN problem is that important information is faded away. It uses a back-propagation algorithm in order to replicate its results through layers. Although their main purpose is to learn long-term dependencies, theoretical and empirical evidence shows that it is difficult to learn to store information for very long [2]. For example, it would be harder for an RNN to predict the next correct word for a 2000 words text than for a 10 words sentence. This situation is called vanishing gradient problem.

LSTM networks are specifically designed to learn long-term dependencies and are capable of overcoming the previously inherent problems of RNNs, i.e., vanishing and exploding gradients [8]. This technique was first proposed by Hochreiter and Schmidhuber (1997) [11] and aimed solving the vanishing gradient problem. It does so by keeping the error flow constant through special units called "gates" which allows for weights adjustments as well as truncation of the gradient when its information is not necessary [23]. Thus, the vanishing problem is solved and the RNN doesn't lose the important data through time.

4   Results

The results are presented in three stages. First, information about the data is presented. The second step consists in using the data set in order to cluster the stocks by its similarity. Lastly, by using the groups created in the last part, the LSTM is trained, and the predictions are made. The results are discussed in Sect. 5.

After analyzing the download data set, it was possible to observe that the period between 01/01/2008 and 29/03/2018 contains the biggest portion of the data. The result of the extraction is shown in Table 1, both exchanges had their size data reduced drastically.

Table 1. Representation of the dataset before and after extraction

Exchange	Number of files	Number of extracted files
NYSE	3795	1225
NASDAQ	3066	804

4.1 Clustering

In order to illustrate the results obtained in the last step, a correlation matrix was plotted and is presented in Fig. 1. For each stock, in each stock exchange, the correlation between all of them was measured. It is possible to confirm that the main diagonal equals to zero and the rest of the matrix is between −1 and 1. Therefore, it is possible to use this result to group similar stocks.

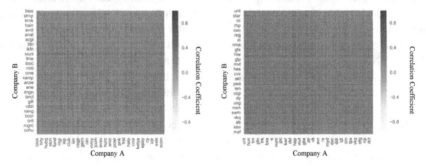

Fig. 1. Correlation Matrix. a) NASDAQ (Left); b) NYSE (Right).

The elbow method was used to obtain the optimal number of clusters. It consists in plotting a graph, where it is possible to check the number of clusters versus the generated error, in this case it was utilized the Within-Cluster Sum of Squares. It is important to emphasize that we cannot just choose the biggest number of clusters, with the minimum error value, because it would cause an over-fit.

Using the information generated by K-Means algorithm, the data was grouped using simple average, which means that for each day the values of all group components where summed and then divided by the number of components, it was done for every day since 01/01/2008 in both exchanges. The Table 2 shows the number of files in the beginning of the process, and the number of groups for each exchange. By doing that, instead of analyzing 2029 predictions, it will be necessary to evaluate just 40.

Table 2. Number of groups

Exchange	Files	Files after extraction	Number of groups
NYSE	3795	1225	20
NASDAQ	3066	804	20

4.2 Predicting Values

The data set was divided in 80% for training and 20% for tests. Each group of stocks was trained and tested separately. A LSTM network was created in order to check if values generated from clusters could be predicted. The network was developed using accuracy as a metric and the mean squared error (MSE) as the loss function. Some of the predicted values will be shown next.

First, we must evaluate the metrics generated. Table 3 shows all the values sorted by their respective error for both exchanges. We can check that the error is small, but also, the accuracy is really small. In Fig. 2 (a) it is possible to confirm that the error decreased over the time but - as we can check in Fig. 2 (b) - the accuracy of the predictions is not so good. The predicted values (represented by a blue line) are really close to the real values (orange) but not on top of each other. Thus, the predicted values are not the ideal ones, but the network produced potential results.

Table 3. Results for both exchanges

Exchange	Group	MSE error	Exchange	Group	MSE error
Nasdaq	Group_11	0.000241121	NYSE	Group_9	0.000353639
Nasdaq	Group_10	0.001105024	NYSE	Group_17	0.000464707
Nasdaq	Group_0	0.001893583	NYSE	Group_15	0.000620887
Nasdaq	Group_5	0.002498377	NYSE	Group_12	0.000789169
Nasdaq	Group_17	0.002949402	NYSE	Group_10	0.002226865
Nasdaq	Group_3	0.003049409	NYSE	Group_19	0.002670706
Nasdaq	Group_16	0.003102871	NYSE	Group_3	0.003475358
Nasdaq	Group_2	0.003516029	NYSE	Group_5	0.003642246
Nasdaq	Group_9	0.004704485	NYSE	Group_4	0.004226033
Nasdaq	Group_19	0.004859025	NYSE	Group_7	0.00573763
Nasdaq	Group_6	0.005061181	NYSE	Group_13	0.005756983
Nasdaq	Group_4	0.005986108	NYSE	Group_14	0.010200528
Nasdaq	Group_8	0.009971577	NYSE	Group_11	0.011839856
Nasdaq	Group_14	0.011123924	NYSE	Group_1	0.012058058
Nasdaq	Group_1	0.016487216	NYSE	Group_18	0.013770473
Nasdaq	Group_18	0.034465782	NYSE	Group_16	0.049443774
Nasdaq	Group_7	0.054812843	NYSE	Group_6	0.055835894
Nasdaq	Group_15	0.066097262	NYSE	Group_8	0.063193569
Nasdaq	Group_12	0.107150873	NYSE	Group_2	0.129570389
Nasdaq	Group_13	0.145795589	NYSE	Group_0	0.165804342

The same behavior can be observed in Fig. 3 for the NYSE Exchange. In Fig. 3 (a) it is possible to observe how the error decreases, but the predictions are not on top of the real values (Fig. 3 (b)). We can also observe that as the neural network tries to predict one value at a time, we can observe that a mistake in the beginning of the process influences the rest of the prediction.

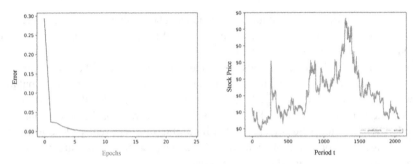

Fig. 2. Results generated by the network for the Group 9 for the NASDAQ Exchange. a) Error vs. Epochs (Left); b) Prediction made by the LSTM (Right). (Color figure online)

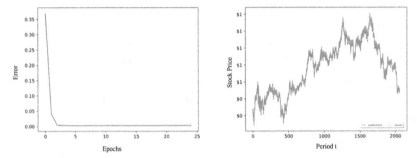

Fig. 3. Results generated by the network for the Group 9 for the NYSE Exchange. a) Error vs. Epochs (Left); b) Prediction made by the LSTM (Right).

5 Discussion

We have shown how two different machine learning techniques can be used in pursuit of a single goal. The raw data was organized, and important information was extracted. The remaining stocks were clustered by their correlation and groups of stocks were generated using K-Means algorithm. Finally, a LSTM network was applied in the data set with the purpose of predicting the value of a group of similar stocks. The process was done successfully, and it proves that stocks can be clustered using K-Means algorithms and that the network prediction capacity did not suffer any impact.

These results can have a huge impact for investors and portfolio managers. By using the proposed techniques, it is possible to buy stocks that have the same behavior. Thus, less time and less computational power will be necessary in order to evaluate a whole group of companies.

Potential directions of future work include analyzing the impact of different similarity measures and how it can influence the process of clustering stocks. Also, when dealing with clustering, it will be necessary to study further options and algorithms that better represent the similarity between stocks. Furthermore, it seems that, even with some error in the beginning of the series, the stock tendency keeps unaltered, so it should be taken in consideration in future work. Additionally, better evaluation metrics should be used in the LSTM network aiming a better accuracy.

References

1. Affonso, F., de Oliveira, F., Dias, T.M.R.: Uma análise dos fatores que influenciam o movimento acionário das empresas petrolíferas. In: Ibero-Latin American Congress on Computational Methods in Engineering (CILAMCE) (2017)
2. Bengio, Y., Simard, P., Frasconi, P.: Learning long-term dependencies with gradient descent is difficult. IEEE Trans. Neural Netw. **5**(2), 157–166 (1994)
3. Bini, B.S., Mathew, T.: Clustering and regression techniques for stock prediction. Procedia Technol. **24**, 1248–1255 (2016)
4. Cavalcante, R.C., Brasileiro, R.C., Souza, V.L., Nobrega, J.P., Oliveira, A.L.: Computational intelligence and financial markets: a survey and future directions. Expert Syst. Appl. **55**, 194–211 (2016)
5. Chollet, F.: Keras (2015). https://keras.io
6. Chong, E., Han, C., Park, F.C.: Deep learning networks for stock market analysis and prediction: methodology, data representations, and case studies. Expert Syst. Appl. **83**, 187–205 (2017)
7. Filho, D.B.F., Júnior, J.A.D.S.: Desvendando os mistérios do coeficiente de correlacão de pearson (r). Universidade Federal de Pernambuco (2009)
8. Fischer, T., Krauss, C.: Deep learning with long short-term memory networks for financial market predictions. Eur. J. Oper. Res. **270**(2), 654–669 (2018)
9. Gan, G., Ma, C., Wu, J.: Data Clustering: Theory, Algorithms, and Applications, vol. 20. SIAM (2007)
10. Gerlein, E.A., McGinnity, M., Belatreche, A., Coleman, S.: Evaluating machine learning classification for financial trading: an empirical approach. Expert Syst. Appl. **54**, 193–207 (2016)
11. Hochreiter, S., Schmidhuber, J.: Long short-term memory. Neural Comput. **9**(8), 1735–1780 (1997)
12. Jung, S.S., Chang, W.: Clustering stocks using partial correlation coefficients. Phys. A: Stat. Mech. Appl. **462**, 410–420 (2016)
13. Ketchen, D.J., Shook, C.L.: The application of cluster analysis in strategic management research: an analysis and critique. Strat. Manag. J. **17**(6), 441–458 (1996)
14. Krauss, C., Do, X.A., Huck, N.: Deep neural networks, gradient-boosted trees, random forests: statistical arbitrage on the S&P 500. Eur. J. Oper. Res. **259**(2), 689–702 (2017)
15. Kumar, B.S., Ravi, V.: A survey of the applications of text mining in financial domain. Knowl.-Based Syst. **114**, 128–147 (2016)
16. Längkvist, M., Karlsson, L., Loutfi, A.: A review of unsupervised feature learning and deep learning for time-series modeling. Pattern Recognit. Lett. **42**, 11–24 (2014)
17. LeCun, Y., Bengio, Y., Hinton, G.: Deep learning. Nature **521**(7553), 436 (2015)
18. Li, Y., Jiang, W., Yang, L., Wu, T.: On neural networks and learning systems for business computing. Neurocomputing **275**, 1150–1159 (2018)
19. Liu, Y., Li, Z., Xiong, H., Gao, X., Wu, J.: Understanding of internal clustering validation measures. In: ICDM, pp. 911–916 (2010)
20. Mirkin, B.G.: Mathematical Classification and Clustering. Kluwer Academic Publishing, Dordrecht (1996)
21. Momeni, M., Mohseni, M., Soofi, M.: Clustering stock market companies via k-means algorithm. Kuwait Chap. Arab. J. Bus. Manag. Rev. **4**(5), 1 (2015)

22. Nanda, S.R., Mahanty, B., Tiwari, M.K.: Clustering Indian stock market data for portfolio management. Expert Syst. Appl. **37**(12), 8793–8798 (2010)

23. Nelson, D.M., Pereira, A.C., de Oliveira, R.A.: Stock market's price movement prediction with LSTM neural networks. In: 2017 International Joint Conference on Neural Networks (IJCNN), pp. 1419–1426. IEEE (2017)

24. Python Software Foundation: Python 3.5.5 documentation (2018). https://docs.python.org/3.5/

25. Qiu, M., Song, Y., Akagi, F.: Application of artificial neural network for the prediction of stock market returns: the case of the Japanese stock market. Chaos, Solitons Fractals **85**, 1–7 (2016)

26. Zhang, J., Cui, S., Xu, Y., Li, Q., Li, T.: A novel data-driven stock price trend prediction system. Expert Syst. Appl. **97**, 60–69 (2018)

Concepts in Topics. Using Word Embeddings to Leverage the Outcomes of Topic Modeling for the Exploration of Digitized Archival Collections

Mathias Coeckelbergs[(✉)] and Seth Van Hooland

Université libre de Bruxelles, Brussels, Belgium
{mcoeckel,svhoolan}@ulb.ac.be
http://mastic.ulb.ac.be/

Abstract. Within the field of Digital Humanities, unsupervised machine learning techniques such as topic modeling have gained a lot of attention over the last years to explore vast volumes of non-structured textual data. Even if this technique is useful to capture recurring themes across document sets which have no metadata, the interpretation of topics has been consistently highlighted in the literature as problematic. This paper proposes a novel method based on Word Embeddings to facilitate the interpretation of terms which constituted a topic, allowing to discern different concepts automatically within a topic. In order to demonstrate this method, the paper uses the "Cabinet Papers" held and digitised by the The National Archives (TNA) of the United Kingdom (UK). After a discussion of our results, based on coherence measures, we provide details of how we can linguistically interpret these results.

Keywords: Topic modeling · Word embeddings · Document classification · Information retrieval

1 Introduction

The central bottleneck in the current topic modeling practice is how to interpret the developed models. As evidenced early by [7], it is difficult to compare different models, even if some methods such as topic intrusion have become well-known. Although these papers provide an important basis for the evaluation procedure, they remain stuck with the large amount of seemingly unavoidable subjectivity in the evaluation of topic models. Next to these interpretational tasks, a series of coherence measures have been developed to evaluate the intra-topic coherence of top terms. As we will discover later in this paper, these methods measure the overall coherence, but do not evaluate the local context of these words.

   On the other hand, word embeddings have proven their worth to derive semantic information from a given corpus, for which tasks such as word analogies

© ICST Institute for Computer Sciences, Social Informatics and Telecommunications Engineering 2020
Published by Springer Nature Switzerland AG 2020. All Rights Reserved
R. Mugnaini (Ed.): DIONE 2020, LNICST 319, pp. 41–52, 2020.
https://doi.org/10.1007/978-3-030-50072-6_4

receive salient scores. Although word embeddings can also be used for document classification, the results focus on the discovery of concepts, understood as combinations of words which are strongly related on a semantic level. The usefulness of this feature has been amply discussed in the literature, but the question as to what extent concepts occur together, remains difficult to approach using only word embeddings.

This article present a novel method to combine word embeddings and topic models in order to compare their unique way of modeling a document collection. Both methods reside within their own research space, respectively neural network models for word embeddings, and information retrieval models for topic modeling. Whereas the former are more interested in modeling language in and of itself, and hence are more interested in the meaning of texts, the latter focus on the retrieval of the most relevant documents. In this article, we seek to answer two research questions. Firstly, we address the way in which Topic Modeling can be used in a specific archival context, in particular a large subset of digitized archival holdings of the National Archives of London (TNA)[1]. [10] demonstrated how to extract re-occurring themes via topic modeling (Latent Dirichlet Allocation). This is particularly useful to circumvent the problem of limited accessibility of digitized archival collections due to the minimal metadata which are available, hence severely limiting the possibilities in which historians and other interested people can interact with these documents.

Secondly, we wish to assess the viability of pre-trained word embeddings on a very large corpus (more than one billion tokens) to serve as a model of an entire (synchronic) language, which can help in semantic problems such as text summarization and query expansion. In this way we can use topic modeling to discover the topics important to the document collection, whereas word embeddings are used to discover concepts inherent in those topics. In other words, we compare the global context modeled by topic modeling with the local context described by word embeddings. By using these two techniques together, we are able to assess the concepts found throughout the document collection, as well as the way in which they co-occur. The remainder of this article will be structured in three parts. In the second sections, an overview of existing work on both topic models and word embeddings is presented, while putting an emphasis on the current open-ended questions. In the third section, then, we propose our novel methodology for combining both techniques, and describe the main results based on some examples. The last section discusses the ways in which we can expand our research.

2   Brief Overview of Topic Modeling and Word Embeddings

The use of vector space models in natural language processing (NLP) has proven very useful since its inception in the 1990s. Since [8] released their paper on

---

[1] https://www.nationalarchives.gov.uk/cabinetpapers/.

Latent Semantic Indexing (LSI), these methods have been applied in various domains of NLP, such as text summarisation, document classification and sentiment analysis. Their method is capable of finding (semantic) similarities between terms, starting from the word-document co-occurrence matrix, and then continue to use singular value decomposition to achieve dimensionality reduction. From a historical standpoint, this approach could be seen as topic modeling avant-la-lettre, because it results in the words of the original matrix to be clustered together in lower dimension. However, as we will see in the following subsection, the term topic modeling is currently understood to refer to the set of algorithms which use a probabilistic method to achieve this clustering. After this introduction to topic modeling, we will see how the use of vector spaces arises again within the context of word embeddings.

2.1   Topic Modeling

The use of the term 'topic model' has seen a semantic shift in its short but intense lifetime. At the moment, the term 'topic model' is near synonymous with the -by far- most widely used algorithm, namely Latent Dirichlet Allocation (LDA). This generative probabilistic model, published in the seminal 2003 paper by [3], clusters key words extracted from a document collection together in such a way, that they can serve as a source to generate the document collection. This model presumes that topics are hidden in the document collection, which can be rendered explicitly by the aforementioned clusters. Before this seminal paper, other methods of clustering key terms together were developed, which could also be marked as a type of topic model, such as the above mentioned LSI technique by [8]. After the seminal publication of [3], other research has vastly expanded on the conception of topic models. The most important works on the algorithmic evolution include hierarchical LDA [4]), which allows to hierarchically structure the extracted topics, and correlated LDA [5], which models the correlation of different topics. Next to these works which bear witness to the usefulness and wide applicative potential of these models, other voices have underlined some challenges. Two main ones arise within the literature, which both put an emphasis on the inherent subjective aspects of topic models. Firstly, the user needs to make choices concerning the amount of topics the algorithm needs to end within the document collection, as well as to assess whether the corpus under scrutiny is ready to be modeled. As has been described by other researchers, good decisions on both aspects can only be done after trial and error experimentation with the corpus and the algorithm(s), which then afterwards have to be assessed for their salience and applicability. This result brings us directly to the second problematic aspect of topic models, namely their interpretation. As [7] have indicated, it is difficult to present objective standards to monitor which interpretations of the topic model are valid and which not. Through their proposed methods of topic intrusion and word intrusion, they provide a measure of the stability of the topic through semantic relatedness.

Going further in this same direction, [10] built on this idea within a multilingual setting, although they still confess to the a priori interpretational difficulty of topic models. This difficulty arises from the fact that it is psychologically attractive for humans to give a meaningful interpretation to a list of words they are presented. Even though given several clear cases -which often are cherry-picked-, we can see that a clear interpretation is sometimes allowed, but it is difficult to discern where the grey area of interpretation is located. This results from an interpretational difficulty inherent in topic models, namely that we would like to represent concepts hidden within the text. Although we know that the clusters of keywords are merely a representation of their occurrence within the document collection, we expect them to correspond to clear-cut concepts. This is due to the distributional hypothesis within the field of linguistic semantics, which states that the meaning of a word is determined by the company it keeps. Expressed differently, this hypothesis understands words which can occur in similar contexts to have a semantic relatedness. In practice we see that it can be the case that topics express concepts, for example the concept of DNA, biology, in the famous example from [3]. However, looking from a practical perspective, we see that topics in the grey zone of interpretation do not allow such easy identification with concepts, but rather as a combination of two or more concepts. This is because topic models derive from an information retrieval context, focussing on co-occurrence of terms, rather than a computational linguistic background which focusses on meaning. Of course, it is also attested that noise can be present, even in well-trained models, due to their inherent probabilistic nature. In the next section we explore how word embeddings are used to represent concepts in a local context.

2.2   Word Embeddings

In contrast to the topic models, which rely on probabilistic measures, word embeddings rely on the vector space model. The term word embeddings was first coined by [2] as a constitutive part of their neural language model. The term was made popular ten years later by the seminal paper of [11], in which they describe Word2Vec, an online, freely available toolkit to either train word embeddings on a corpus, or to use their pre-trained word vectors. Given the increasing attention for word embeddings in the NLP community, a year later GloVe was introduced by [13], a comparable toolkit to Word2Vec, which takes a slightly different approach and is the main competitor of Word2Vec. Since these two models, Word2Vec and GloVe are responsible for the current high output of NLP work based on word embeddings, we will limit our presentation to these two, since they diverge on a few key points, necessary for the remainder of this article.

The two models correspond to each other in some general aspects. They both learn a vectorial representation from the co-occurrence of words with the broader context in which they appear. In this way, both models represent geometrical encodings of the words from a corpus. The main difference between both is that Word2Vec is a predictive model, while GloVe is count-based. More specifically, this means that Word2Vec uses the vectorial representation to minimise a loss function predicting the target words from the context words. The lower the function, the better the predictive power of the model. Word2Vec uses a feed-forward neural network for this task, while using Stochastic Gradient Descent for smoothing. In particular, every feed-forward neural network turns the input vocabulary into word embeddings, found in the weights of the first layer, which hence is also known as the embedding layer. [11] describe two architectures for learning these word embeddings, namely Continuous Bag-of-Words (CBOW) and skip-grams. These CBOW architecture, as the name already implies, uses continuous representations of the documents, and hence does not pay attention to the word order. The model starts from different input words, forming the context, and from this it aims to derive the probability for a specific word. The skip-gram architecture does the exact opposite, meaning it starts from a specific word and tries to derive probabilities for the context words. In contrast to this neural network approach of Word2Vec, GloVe starts of from the co-occurrence counts, represented through a word-context matrix. This is a high-dimensional matrix, both in rows and columns, which needs to be reduced in dimensionality to be able to derive semantic information. For GloVe, this is done by an objective function with weighted least squares with the differences between two words encoded as vector differences. This approach allows GloVe to make the relationships between words its inherent point of departure, whereas for Word2Vec it rather is a by-product.

Word embeddings are strongly related to distributional semantics models (DSM), and as a matter of fact outperform them. In fact, given that the distributional hypothesis states that the meaning of a word is determined by the company it keeps [9], we could infer that this is precisely what the above models also try to convey. The difference between DSMs and word embeddings is that the former keep track of contextual information by registering co-occurrences and counting them, whereas the latter use the same information, but use this mainly to predict surrounding words. As is shown by [1], the latter models nearly always outperform the former models. While DSMs are clearly count models, and Word2Vec clearly is a predict model, outperforming the former due to its neural architecture, the nature of the GloVe model is contested. Since it is factorising a co-occurrence matrix which links word to their context, it is based on the counts of words, being very close to traditional vector-based methods such as Latent Semantic Indexing. From this perspective, it should be classified as a count model, while [1] consider it a predict model as it also tries to minimise a loss function. Within the context of this paper, we seek to use the word embeddings as useful vectorial representations of the language as a whole, which then allows to estimate the semantic relatedness of terms found in the same topic.

2.3   Syntagmatic and Paradigmatic Axes from Structural Linguistics

As we have established in the previous sections, we obtain a global and local perspective on textual meaning by respectively using topic modeling or word embeddings. Since word embeddings trained on a large dataset can be interpreted as representative of the language as a whole, as in the GoogleNews dataset, they can improve the coherence found by the document-level top-ranked words of topic modeling. On a linguistic level, the difference between these results have been interpreted to accord with Ferdinand De Saussure's seminal distinction between syntagmatic and paradigmatic contexts. The syntagmatic axis looks at the phrase-level at the direct context of a word under scrutiny. This accords well with the word embeddings approach, which takes a context window into account for every word. The paradigmatic axis on the other hand looks at the replaceability of each word in context. This accords well with topic modeling, since the same word can occur in different topic distributions if it can be injected in different contexts. This is a good test of polysemy. Explained differently, the syntagmatic axis for the word 'man' will register typical contextual words such as 'cried' and 'the', whereas the paradigmatic axis measures which words can be placed in the same contexts as 'man'. In this case, 'girl' is an example of a paradigmatical equivalent to 'man'. Figure 1 shows a simple visual example of the syntagmatic-paradigmatic axes.

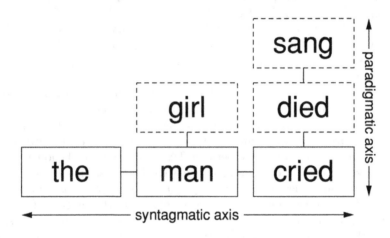

Fig. 1. Example of Syntagmatic-paradigmatic axes [6]

Interpreting the results of the previous section, this means that the keywords occurring within a topic have a high likelihood of being interchangeable, as the paradigmatic axis indicates. Although they have a high rate of interchangeability, this does not necessarily mean that the local contexts in which these words generally appear, would also be similar. This can be measured by word embeddings on a big dataset, to measure the relevance of local contexts within the English language. Reordering the topic modeling results based on the vectorial representation of word embeddings thus means that we evaluate the top paradigmatically similar terms based on their syntagmatic similarity. In our results section we will provide a concrete example of how to interpret the output of the topic modelling and word embeddings step along the syntagmatic-paradigmatic axes.

3 Mobilizing Word Embeddings to Discern Concepts Within Topics - A Novel Methodology

In this section we develop our methodology to use (pre-trained) Word2Vec word embeddings as vectorial representation to determine the semantic relatedness of keywords clustered together as a topic. This method should allow us to decrease the subjectivity of interpretation of topics, since it is easier to determine the underlying concepts. In this way, we could state that the topics are mainly used to discern what the documents are talking about -their global context-, whereas word embeddings are used to find semantic structure -the local context- in the probabilistic outcome of topic modeling. To be clear, we do not propose to use word embeddings as a rival classification method to topic modeling. As we have shown above, this viability has already been documented within the literature. As a connection between concrete documents and a specific vocabulary we use topic modeling, and propose word embeddings to help answer one of the inherently subjective questions of topic modeling techniques in general, namely how many topics to establish.

As we have learned from the literature overview, topic modeling helps to explore themes from a corpus under scrutiny, whereas (pre-trained) word embeddings can be seen as a general, vectorial representation of the language itself. Bringing together both sources of information allows the information gathered by topic modeling to be refined, allowing us to estimate the content of the extracted topics more concretely. Within this paper, we do not wish to evaluate the influence of different topic distributions, that is, to measure the influence of the choice of amount of topics to be retrieved on their internal coherence. Nevertheless, the methodology followed here can be re-used to other topic distributions, which then in turn can be compared. Different questions can be answered in this regard, depending on the research question leading to the conception of the topic modeling. The most prominent examples include the search for the topic distribution which most saliently models a certain concept. For this step, a reconciliation with a controlled vocabulary such as a thesaurus has to be made, but this step is out of the scope of the current article. Within the methodology developed, we divide the approach in separate sections on respectively topic modeling and word embeddings. By doing so, individual parts of the methodology can be used again, for example by exchanging the topic modeling approach by another clustering method, or by testing the coherence of the clusters by another method than word embeddings.

Although the methodological problems of producing salient topic models are cumbersome, we do not wish to go into the results achieved through different parameters entered in the algorithm. The main purpose of this article is to discuss how the semantic relatedness of words in a topic cluster can be evaluated, and how in turn different models can be assessed for their salience concerning a specific query. For this purpose, it suffices to outline the general methodology for modeling our corpus. For the purpose of creating the topic model, we used the machine learning package Gensim in Python, which also allows other machine learning applications such as document classification or information extraction.

We have to make certain decisions concerning stop word removal and the amount of topics, two aspects we will explain now. The stop word list contains a list of words which have no semantic contribution to the model, and hence can be deleted from consideration. Some words deserve a place within this list without hesitation, and can be easily accessed online, where several general-purpose stop word lists are readily available. These lists include such often attested, semantically vacuous words, such as interjections, conjunctions, and personal pronouns. Since the purpose of the current article is to propose a method to automatically identify the semantic coherence of the words in a topic, we will not go further than this standard list in appropriating it for our corpus. A further improvement could include a type of iterative selection, where firstly a topic model is created with only the standard stop words deleted, after which specific words, which either appear as outliers within the dataset, given the general scope of the topics under scrutiny, or words such a broad meaning that they do not have any delimiting power. For example, if our corpus is very strongly economic in nature, words such as 'economics', 'investment' and 'payment' are so widespread that they together constitute an important topic for each document. The number of topics which needs to be selected is a more difficult aspect than the selection of stop words. As we have mentioned in our state of the art on topic modeling, there is no specific standard to derive the amount of topics which should be extracted given a certain corpus size, language of content. The only general advice or rule which can be given is that the number should be such that it produces coherent topics, which in turn also is a concept which is difficult to define. It can be said in general that there is a middle zone in which the 'ideal' number of topics can be found, between having only topics with the most famous words, which do not indicate any semantic distinction between documents or topics when the number of topics selected is too low, or having an overproduction of nearly the same topics, with several noise terms when the number is too high. The reason why 'ideal' is placed between quotes is that in general practice, an approximation seems to be used, whereas only whole- number amounts of topics are tested, leaving the appropriateness of 100 topics rather than 97 or 102 unaccounted for. While admitting our subjective bias, we estimate that the amount of 50 topics gives a workable basis for the continuing questions of the article. This assessment is based on hyperparameter optimization, where the results are interpreted by the authors of this article.

Word embeddings deliver a vectorial representation for every word present the training corpus, and when trained on a representative corpus of sufficient size, these vectors can represent an entire natural language. This can be used subsequently for the representation of each keyword in the topic cluster. As was discussed in the state of the art, Word2Vec generally outperforms GloVe and was therefore chosen in the context for this paper. Also, preference has been given to use pre-trained word vectors, as we wish to underline their general-purpose use, which we apply here to topic modeling.

4   Results of the Co-occurrence of Terms

Word embeddings offer the possibility to select the best topic distribution, but this does not yet give us a clear vision of what we can find inside these topics precisely. From the LDA algorithm we derive that topics are formed according to the co-occurrence of certain words. Hence, words which are found inside the same topic have a high tendency to occur closely to each other in the document collection. From the general outline of topic modeling research, we find that a topic model should be able to identify what a text is about. In an ideal setting, users can identify the collection of keywords grouped by the algorithm to constitute a topic with a human concept, hence allowing for easy interpretation of the topics. For example, in the underlying example of the fifth topic, we see how the words *oil price commission energy coal company market supply demand opec increase percent production scheme index industry countries spot world stocks* have a high tendency of occurring together throughout the documents. Applying word embeddings to these words, and re-arranging them from the word most similar to the other words all the way to the word least similar to the other ones, we find the following: *market price demand supply industry company stocks production opec index percent increase oil energy coal country world commission scheme spot*. Throughout the experience of dealing with topic models, several researchers have pointed out that, although some clear-cut cases are available, the lion's share of clusters of key words lead to interpretational difficulties. Hence, we could state that it is a rare case to find that a topic corresponds to a human concept, and that in general we find multiple concepts represented in the same topic. In the above example we could state that 'market price demand supply industry company stocks' forms a concept of economy, whereas 'oil energy coal' forms a concept of energy. Taking a bigger window than the twenty most relevant terms for the topic might reveal other terms which word embeddings would group belonging to either of these two concepts. Of course it is difficult to objectively score whether the re-arrangement of topical keywords based on their vectorial representation. Showing the coherence according to three different metrics of three randomly chosen topics, before and after re-arrangement, shows that in general improvement is attested. This means that indexation of the documents will be more coherent and relevant if the top term after filtering by word embeddings information is performed than before. The three coherence measures are based on the Palmetto Toolbox, where CV is based on a sliding window, a one-set segmentation of the top words and an indirect confirmation measure that uses normalized pointwise mutual information (NPMI) and the cosinus similarity., CP on a sliding window, a one-preceding segmentation of the top words and the confirmation measure of Fitelson's coherence, and CUCI on a sliding window and the pointwise mutual information (PMI) of all word pairs of the given top words. CV and CP are based on [14], CUCI on [12].

Topic number	Measure	Before Re-arrangement	After Re-arrangement
3	CV	0.2608	0.2990
3	CP	−0.2748	−0.0776
3	CUCI	−0.5651	0.1757
18	CV	0.3872	0.4011
18	CP	0.4096	0.5233
18	CUCI	1.1214	1.4375
39	CV	0.3078	0.3479
39	CP	−0.3573	−0.2385
39	CUCI	1.0682	1.2685

These results show that re-arranging the top keywords from the topic modeling results using word embeddings gives an increase in the score across these three measures. We have selected three topics at random, in order to overload the table with numbers.

5   Linguistic Interpretation of the Results

We have shown how a local context window of word embeddings can improve the coherence of the global topic modeling results. As we have indicated in our overview section, these results can be interpreted using the linguistic distinction between syntagmatic and paradigmatic axes. Interpreting the results of the previous section, this means that the keywords occurring within a topic have a high likelihood of being interchangeable, as the paradigmatic axis indicates. Although they have a high rate of interchangeability, this does not necessarily mean that the local contexts in which these words generally appear, would also be similar. This can be measured by word embeddings on a big dataset, to measure the relevance of local contexts within the English language. Reordering the topic modeling results based on the vectorial representation of word embeddings thus means that we evaluate the top paradigmatically similar terms based on their syntagmatic similarity.

Going further, we can also compare rankings based on both models. Using word embeddings, the local context for every word can be compared by ranking the remainder of the vocabulary according to vector similarity to the word under scrutiny. For topic models, we need to find all occurrences of a given word among the different topic distributions. Each distribution will give a score for every word in the vocabulary, allowing us to find the mean value for every term with respect to a word under scrutiny. This ranking using the topic modeling results is more difficult, because the algorithm does not provide results on the word-level, only on the topic-level and document-level. When both rankings, the one of topic modeling and word embeddings, are similar, this means that paradigmatic and syntagmatic axis overlap. In other words, this means that the document-level word relationships are similar to the ones based on the smaller context window of word embeddings. In such case, the word under scrutiny is used in an expected way.

Taking the word 'electricity' for example, we find that based on the paradigmatic axis, 'fuel' is the most similar word. This means that, given the corpus under investigation, these two words have highly similar contexts. As two of the most important energy resources, we can imagine that indeed on a global level many words have the same likelihood for both words. Based on the word embeddings, 'fuel' is only ranked on the eleventh place for similarity to 'electricity'. This means that in the average sentence, both of these words do not have a very high likelihood of occurring together. The most similar local level word for 'electricity' is 'megawatts', and for fuel 'gasoline'. These are expected results, since megawatts and gasoline are very likely to be in sentences which also have respectively 'electricity' and 'fuel' in them (Fig. 2).

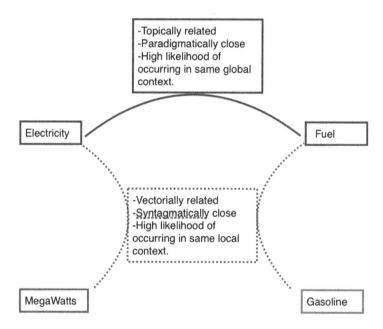

Fig. 2. Schematic Rendering of the relationship between Topically and Vectorially related Words

6   Conclusions and Future Work

In this article, we have sought to explore to what extent the topic modeling and word embedding models show comparable results. We have explained how the former models the global context of a document collection, whereas the latter models the local context. Using local information on global data improves the coherence of the top keywords from topic modeling. This leads to more salient words for indexing or query expansion models. Next to this fully automatic pipeline to improve indexing using topic modeling, we have also shown how the results of both models can be interpreted linguistically, using the concept of syntagmatic and paradigmatic axes from Ferdinand de Saussure. This interpretation has allowed us to interpret the local and global differences for a concrete example from the corpus. Future work will consist of comparing different architectures within topic models and word embedding models. In this paper we have opted to only use the most well-known incarnation of both, respectively Latent Dirichlet Allocation and Word2Vec.

References

1. Baroni, M., Dinu, G., Kruszewski, G.: Don't count, predict! A systematic comparison of context-counting vs. context-predicting semantic vectors. In: Proceedings of the 52nd Annual Meeting of the Association for Computational Linguistics, pp. 238–247 (2014)
2. Bengio, Y., Ducharme, R., Vincent, P., Jauvin, C.: A neural probabilistic language model. J. Mach. Learn. Res. **3**, 1137–1155 (2003)
3. Blei, D.M., Ng, A., Jordan, M.: Latent Dirichlet allocation. J. Mach. Learn. Res. **3**, 993–1022 (2003)
4. Blei, D.M., Griffiths, T.L., Jordan, M., Tenenbaum, J.: Hierarchical topic models and the nested Chinese restaurant process. In: Advances in Neural Information Processing Systems 16 (2004)
5. Blei, D.M., Lafferty, J.D.: Correlated topic models. In: Weiss, Y., Schölkopf, B., Platt, J. (eds.) Advances in Neural Information Processing Systems 18. MIT Press, Cambridge (2006)
6. Chandler, D.: Semiotics: The Basics, 2nd edn. Routledge, London (2007)
7. Chang, J., Boyd-Graber, J., Gerrish, S., Wang, C., Blei, D.M.: Reading tea leaves: how humans interpret topic models. In: Proceedings of the 22nd International Conference on Neural Information Processing Systems, pp. 288–296 (2016)
8. Deerwester, S.C., Dumais, S.T., Landauer, T.K., Harshman, R.: Indexing by latent semantic analysis. J. Am. Soc. Inf. Sci. **41**, 391–407 (1990)
9. Firth, J.R.: Papers in Linguistics 1934–1951. Oxford, London (1957)
10. Hengchen, S., Coeckelbergs, M., Van Hooland, S.: Exploring archives with probabilistic models: topic modeling for the valorization of digitised archives of the European Commission. In: IEEE International Conference on Big Data Workshop on Computational Archival Science, Washington D.C., pp. 3245–3249 (2016)
11. Mikolov, T., Sutskever, I., Chen, K., Corrado, G., Dean, J.: Distributed representations of words and phrases and their compositionality. In: Proceedings of the 26th International Conference on Neural Information Processing Systems 2, pp. 3111–3119 (2013)
12. Newman, D., Lau, J.H., Grieser, K., Baldwin, T.: Automatic evaluation of topic coherence. In: Human Language Technologies: The 2010 Annual Conference of the North American Chapter of the Association for Computational Linguistics, pp. 100–108 (2010)
13. Pennington, J., Socher, R., Manning, C.D.: Glove: global vectors for word representation. In: Proceedings of the Conference on Empirical Methods on Natural Language Processing (EMNLP), pp. 1532–1543 (2014)
14. Röder, M., Both, A., Hinneburg, A.: Exploring the space of topic coherence measures. In: Proceedings of the Eighth International Conference on Web Search and Data Mining, pp. 399–408 (2015)

A New Entity Extraction Model Based on Journalistic Brazilian Portuguese Language to Enhance Named Entity Recognition

Rogerio de Aquino Silva📖, Luana da Silva📖, Moisés Lima Dutra📖,
and Gustavo Medeiros de Araujo[✉]📖

Engineering and Data Science Lab, Federal University of Santa Catarina,
Florianópolis, Brazil
{rogerio.aquino,silva.luana}@posgrad.ufsc.br,
{moises.dutra,gustavo.araujo}@ufsc.br

Abstract. Named Entity Recognition (NER) plays an important role on broad natural language processing applicability. According to the literature, the NER process applied to the English language reaches around 90% of accuracy. However, when applied to Portuguese, this accuracy is at most 83.38%. A wide range of algorithms based on LSTM (Long-Short Term Memory) architecture has being proposed to enhance the NER accuracy. However, a key component to a successful information extraction is the corpora used for NER training. In order to improve the NER in Portuguese language, this paper proposes a methodology for training text corpus based on Portuguese-language journalistic corpora. The Journalistic language has the best adherence to the contemporaneity of the language, since it preserves features such as objectivity, simplicity, impartiality, and is a reference of transmitting the information without ambiguity. The proposed methodology provides a model to extract entities and assess the obtained results with the use of Recurrent Neural Network architectures. At the best of our knowledge, with the proposed methodology, the NER task applied to the Portuguese language overcomes the average accuracy found in the literature, increased from 83.38% to 85.64%. Moreover, the use of this methodology could decrease the computational costs related to the NER processing tasks.

Keywords: Natural Language Processing · Name entity recognition · Entity extraction model · Brazilian Portuguese corpus · Recurrent Neural Networks

1 Introduction

Information retrieval (IR) has emerged from efforts on facilitating large-scale data manipulation [8]. Efforts to IR development still face major challenges when

R. Mugnaini (Ed.): DIONE 2020, LNICST 319, pp. 53–63, 2020.
https://doi.org/10.1007/978-3-030-50072-6_5

it comes to Natural Language Processing (NLP). The extraction of information from Portuguese texts is still an open field of investigation. It is a consequence of the weak results of NLP models for Portuguese language [2].

In addition to the challenge of applying NLP for Portuguese, there is a growing scale of data generated by the increasing number of internet users [7]. It is estimated that at every second thousands of data are generated in a variety of formats such as images, text, videos, and audios.

A study conducted by Data Management Association (Dama) found there are currently over 500 quadrillion of Megabytes of data stored in the digital universe. The study also pointed out that every two years, the production of data doubles, forecasting to reach 350 zettabytes by 2020 [9]. The large majority of the data generated are unstructured data, i.e. text written in natural language. The increasing scale of text production and the low average accuracies obtained when applying NER to Portuguese are issues that can be tackled with modern machine learning techniques, such as artificial neural networks, decision trees, support vector machines, and Bayesian networks.

By using such techniques, it is possible to create models with the ability to extract entities from texts in natural language. In the English language, for example, some models achieve accuracy of 92.6% when using the spaCy[1] framework, and of 91.7% with the ClearNLP[2] framework, which focus on the training of models for entity recognition.

In order for a model to achieve high performance, it is necessary to provide it with a pre-sorted dataset that possesses notations about entities and the grammatical structure. This dataset is known as textual corpus. Usually, the textual corpus is constructed by the aggregation of text excerpts, which is performed by linguists who know the language structure and its properties. However, since each corpus is created for a purpose they do not have always the same structure, so there is no pattern between them.

Moreover, [19] points out that the morphology and syntax of the Portuguese language have their own characteristics. When compared to the English language, which has fewer elements in its grammatical notation – specially in the conjugation of verbs –, those characteristics generate a more complex scenario to deal with.

Language resources are an important feature when developing computational methods to analyze and study languages [3]. In order to build accurate classifiers, there is a need for quality corpora in order to develop and evaluate classifier models [12]. The key problem in this area of research is how to build large natural language corpora enriched with morphosyntactic information [3]. There are just a few available corpora for Portuguese NER, in which the annotated datasets are usually small and/or insufficient for achieving high accuracies for Portuguese NER. Therefore, reproducing and benchmarking the results of previous works is

---

[1] https://spacy.io/.
[2] https://github.com/clearnlp.

not simple due to the variety of possible dataset combinations and the lack of a standardized methodology for training and evaluating [17]. According to [16], in Portuguese the accuracy is around 83.38%, when one uses the HAREM[3] corpus as a training base to be taken in CRF neural networks.

The scope of information is also a matter to be considered. According to [5], a particular community may have habits regarding the use of information, i.e. how to perform its searches and how to organize its new knowledge. These habits affect not only how the text is written but also how some terms are syntactically structured, even when different communities share the same language. When considering the journalistic writing, a style that has the best adherence to the contemporaneity of the language, some of the characteristics of it such as objectivity, simplicity, impartiality, and referential, use to avoid unusual terms because the information must be clearly transmitted to the reader [15]. For this reason, we chose textual corpora based on journalistic texts to create a training base for our NER approach.

Several machine learning techniques allow the creation of models capable of recognizing entities. One of the most used for this purpose is Recurrent Neural Networks (RNN) [11]. RNNs are generally applied to problems where there is a need for pattern recognition that varies over a given series [10]. However, NER is not a simple task. Several categories of named entities are written similarly and appear in similar contexts. Furthermore, the same named entity can be classified into different categories depending on the surrounding context and some entities do not appear even in large training datasets [14]. Due to the fact that reproducibility is hard when the corpus is not standardized, this work aims to create a new corpus for the Portuguese language based on journalistic texts extracted from the CETENFolha[4].

In order to test the corpus proposed in this work, we used a recurrent network variation, called Bidirectional Long-Short Term Memory (Bi-LSTM) [20]. The proposed model has been trained based on language syntax and allows new information to be added to the training, resulting in a better classification of texts in a given domain.

The remainder of the paper is structured as follows. Section 2 provides a summary of some works related to corpus building. Section 3 presents the methodology used to build and validate the proposed corpus. Section 4 presents an evaluation of the proposal. Finally, Sect. 5 provides the conclusions of this paper.

2   Related Work

A large number of proposals aiming to develop new corpora for classifiers can be found in the literature. The corpora can be based on social media data, open data from news feed, websites or even data from corporations [12]. The authors

---

[3] https://www.linguateca.pt/HAREM.
[4] https://www.linguateca.pt/cetenfolha.

of [6] developed a corpus for smart environments. It comprises audio and video materials, as well as robot and apartment reactions, and information taken from sensors and actuators. The data was gathered from 62 volunteers and can be used for training automated robots. As argued by the authors, this data is valuable for further in-depth analyses of people's interactions with devices, ambient intelligence and robots in everyday environments. The difference between our work and the work of [6] is that we focus on text data for providing corpora for NLP tasks.

In [4] is presented a domain specific Question-Answering Corpus (QA-Corpus) built with Portuguese tweets and news articles. While using social media, the authors could gather candidate and reliable answers to possible user questions, making the dataset more real. They used deep learning to match questions and rank candidate answers. Both this paper and [4] attempts to create corpora for Portuguese language, which lacks of good datasets/corpora in order to benchmark results. As opposed to [4], which focuses on Q&A Systems, our work focuses on NLP NER task.

A Czech attempt to create a national corpus was reported by [18]. The authors describe a project to build a larger corpus comprised of Czech texts extracted from web pages. The motivation behind this project lies on the fact that the authors believe that large corpora are essential to modern methods of computational linguistics and natural language processing. The difference between our work and [18] is that our work does not use web pages, but improves existing corpora in order to produce a unique, unified and comparable corpus.

An Arabic strive to create a corpus took place by relying on online published newspapers from different Arabic countries. The authors of [1] created the corpus for improving different researches in Information Retrieval, Machine Translation, and Arabic Language Processing, in general. As well as [1], our paper is an attempt to standardize a big and comparable corpus for Portuguese language processing in several areas of research. Besides, both works use newspapers text to build the dataset/corpus.

The authors of [13], made a corpus for a specific language domain, the legal vocabulary. The corpus contains entities such as *"TEMPO"*, *"JURISPRUDEN-CIA"* and *"LEGISLACAO"*. The created model used neural networks LSTM-CRF and LSTM-CNN, with an accuracy of 90.01%. On the other hand, the study [16] of 2019 use the corpus HAREM in two different scenarios. In the first, the corpus is used in its entirety, with the entities *"PESSOA"*, *"ORGANIZACAO"*, *"LOCAL"*, *"VALOR"*, *"TEMPO"*, *"ABSTRACCAO"*, *"OBRA"*, *"ACONTEC-IMENTO"*, *"COISA"* and *"OUTRO"*. In the second scenario, it was considered just the entities *"PESSOA"*, *"LOCAL"*, *"ORGANIZAÇÃO"*, *"DATA"*, and *"VALOR"*. For the first scenario, the accuracy obtained was 74.91%. In its turn, the second scenario obtained 83.38% accuracy. The models used a variation of a set of neural networks like Recurrent LSTMs and convolutional networks CNNs and framework flair.

3   Proposed Methodology

In this section, we present the methodology used for standardizing Portuguese corpora in order to produce a single corpus to be used by NER tasks. The methodology was developed in four modules, as it is shown in Fig. 1.

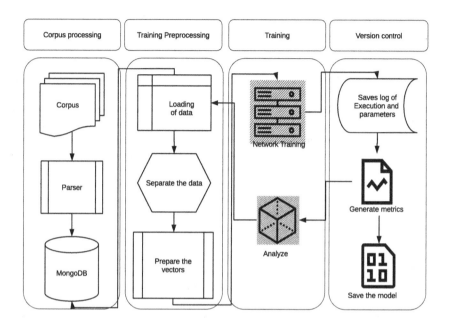

Fig. 1. Structure organization

3.1   Corpus Processing

The first step comprises the parsing of the text to check it for compliance with grammar rules. To create the base corpus, we used the CETEMPublico corpus, which possess a structure in which each line represents a token. A token can be a word, a text fragment, or a punctuation element. A token can be of four distinct types: i) infinitive verb; ii) noun; iii) POS-tag, which indicates a grammatical class; and iv) grammatical detail, which in the case of a verb, would contain the verb itself and its conjugation.

After the parsing, it is necessary to organize the data. Typically, the corpus comprises text files and these files can be massively large, which makes them expensive to process. The process of generating the training base is usually done with multiple datasets at the same time, which can make training which can make training extremely costly computationally. However, the use of a nonrelational database is proposed to mitigate this problem. A structure was created after parsing the information, so that it is organized and inserted into a nonrelational database. This way, other project modules can easily access datasets as needed to generate pre-training.

After parsing and inserting the data into the database, it is necessary to convert the data, as annotations entered into the corpus must meet the demands of non-technology related areas and end up being confusing or not having a pattern on all lines. When training is performed using a neural network we need the data to have the same structure, because if there are incorrect notations the model runs the risk of not being able to generalize and, consequently, not predict correctly. The steps performed were: i) all tokens "PROP" POS-tag were localized, since they are proper names and it is then possible to use them to locate implicit entities; ii) next, the other annotations related to this token that contain $<inst><org><media>$ xml tags were located and converted to ORG, which stands for organization entity; iii) $<hum>$ tags were converted to PER (person entity), and $<civ>$ to LOC (location entity). When this process is finished, the data is sent in the form of three vectors to the training preprocessing module, along with POS-Tag annotations and a list of entities.

3.2   Training Preprocessing

After the training data is received, it is necessary to collect all text excerpts from the vectors, in order to create a unique word vector that will form our final vocabulary known as "word2index". This unique vector is a dictionary containing words and their codes. The same process occurs with the POS-Tag and entity vectors.

Artificial neural networks require data to be sent numerically, so it is necessary to convert the string vectors according to the dictionary, and place each word as an element of the converted list. Each element in this list has also a grammatical class, so the result is a list of tuples. The first element is a number representing a word and the second represents a POS-Tag. The POS-Tag number will be the list vector, where each list represents a text and each element represents a word with its POS-Tag. This input set will be the independent variable in the artificial neural network.

To create the dependent variable, you must convert the entities of each token to numbers. The new list of numeric tokens will be related to the "tag2idx" dictionary. Each entity in this list is related to an entity in the independent variable vector list.

When the token has no localized entity, it is classified as "O". If the token has a localized entity, it is classified with a letter according to its position. If the token is the first token of the entity, it will have the letter "B" (beggin) + "−" + "entity tag". If the token is not the first one, it will have the letter "I" + "−" + "entity tag". The entities tokens defined are:

1. B-LOC: first local entity token, i.e.: "São"
2. I-LOC: remaining location tokens, i.e.: "Paulo"
3. B-ORG: first entity token Organization, i.e.: "Banco"

4. I-ORG: remaining organization tokens, i.e.: "Votorantim"
5. B-PER: first person entity token, i.e.: "Paulo"
6. I-PER: remaining person token, i.e.: "de", "Tarso"
7. O: unclassifiable remaining tokens, i.e.: "Transferência"

Finally, we need to standardize the size of the lists present in the vector, since the artificial neural network needs all vectors to be the same size. For the standardization of vectors, it was found that the maximum size of tokens existing in the texts was of 75 characters. Thus, tokens smaller than 75 characters have been completed with zeros.

3.3   Training

The training module is composed by the parameters responsible for the creation of the artificial neural network such as number of layers, iterations, activation function, and loss function.

The artificial neural network used is called the recurrent neural network (RNN), which is composed of neural units called LSTM (long short-term memory). The LSTM neural unit is a variation of traditional artificial networks. It has the ability to persist information for an arbitrary time. This type of neural unit is widely used in problems where the existence of patterns has a long-term correlation. With RNNs it is possible to connect information from previous memory states to the current one. There are several applications for this type of artificial neural network, such as speech recognition, translation, image captioning, and natural language processing.

In our proposal, we use a LSTM neural unit specification, the bidirectional long short-term memory (BLSTM). As can be seen in Fig. 2, an RNN with BLSTM neural units connects the two-way neural network. Thus, prediction can be performed in both ways, with the aim of maximizing entropy.

Each LSTM unit consists of a cell, an input port, an output port, and a forgetting port that regulate the flow of information. The input port controls new information that can enter in the memory. The forgetting port controls when information should be forgotten by memory, allowing the cell to remember new data. The output port controls the information that leaves the cell. All cells are connected in bidirectional way, by performing forward and backward propagation in each way.

The training artifacts are: i) the binary file model to be reused without the need for retraining; ii) an output metric report against the validation base; iii) the execution time; iv) the parameters used to construct the model; v) the identification code of each text segment of the database, which were used during the process; vi) the history of each iteration of the network along with the loss metrics; vii) the error; viii) a file containing all tokens used in the validation, the true value, and the ones sorted by model.

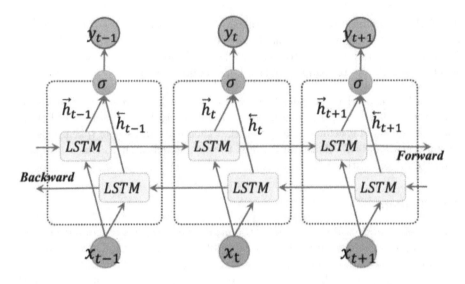

Fig. 2. Recurrent Neural Network with bidirectional long short-term memory

3.4 Version Control

The version control module is responsible for loading the saved data, which includes: the model, the word dictionary, the character dictionary, the tag dictionary, data converted to number, and the conversion to text format again. The main function of this module is to maintain a version history of models created from the training process. Therefore, it is possible to verify the evolution of a model in relation to the accuracy, and thus identify the changes that were significant. Therefore, it is possible to verify the evolution of the model, in relation to the accuracy, and identify the changes that were significant.

4   Methodology Assessments

In this section, we present the assessments of the proposed methodology. The amount of text snippets varied for each entity token, as well as the number of epochs of the training algorithm. As is shown in Fig. 3.

The first evaluation was run with 18001 text snippets and 40 training epochs. The general amount of text snippets was increased and the training epochs was decreased. The main goal was to verify the influence of the size of dataset/corpus for training and the number of epochs required to perform in order to recognize entities.

4.1   Results

In our earlier results, as can be seen in Fig. 4, the accuracy increased for most tokens, from test 1 to test 4:

Test	Tranning Tracks	Epochs	Execution Time (minutes)	Accuracy	Accuracy Isolated
1	18001	40	1334	98,52%	72,65%
2	19001	20	43	98,69%	76,28%
3	28001	20	74	97,23%	83,65%
4	50001	20	147	97,54%	85,64%

Fig. 3. Assessments sets

- B-LOC from 83,44% to 93,36%
- B-ORG from 82,88% to 82,62%
- B-PER from 82,20% to 88,50%
- I-ORG from 25% to 86,61%
- I-PER from 26,53% to 90,91%
- I-LOC from 84,62% to 87,50%

The recognition for I-PER had the highest increased of accuracy around 64%, followed by I-ORG, which the accuracy was increased around 60%. The accuracy for B-LOC was increased around 10%, and B-PER and I-LOC accuracies were increased by around 6% and 3%, respectively. The exception was for recognizing B-ORG entities, to which the accuracy has remained largely the same, with a slight decrease of less than 1%. Regarding the corpus chosen for the tests, the I-LOC entity token was not presented nor in test 1, neither in test 2.

Moreover, the O entity token had also largely the same accuracy during the tests. This kind of entity means represents unclassified token, i.e., in the end, the proposed methodology possess a high accuracy in differentiating words that are not named entities.

In addition, we calculated the isolated accuracy, which means overall correct recognition for all entities, excluding entity O. The isolated accuracy was increased from 72,65% to 85,64%, as the amount snippets was increased, and the total accuracy was kept nearly the same, as it can be seen in Fig. 5. Furthermore, by decreasing the epochs in the training algorithm, the execution time also decreased, as shown in Fig. 3. Most of cloud services capable of processing this kind of training are paid, consequently, the reduction of time and the consumption of computational resources are relevant points to be considered.

5   Final Remarks and Future Works

The preliminary results are promising. It has been proved that regardless of the language domain of the corpus, text parsing is possible. By means of the chosen machine learning algorithm, it was possible to create models that have assertiveness relatively close to those currently obtained by English language-based models.

The biggest challenge is in relation to the annotations, because during the research we found several corpora in Portuguese, mostly incorrect, incomplete or, when complete, composed of insufficient expressive data.

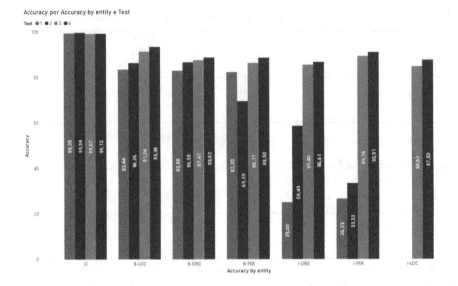

Fig. 4. Accuracy by entity token

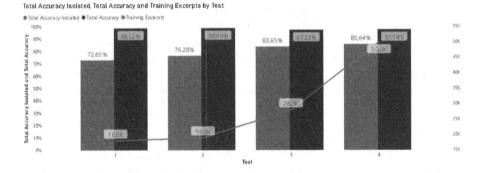

Fig. 5. Total and isolated accuracy

Despite the challenges encountered during this work, it can be said that the overall goal was achieved, since it was possible to create a methodology that was able to extract entities with results close to other results in other languages, such as English, which possess less morphological and syntactic complexity.

The next steps for this research will firstly increase the number of snippets to cover all entities during all test iterations. In the end, we intend to build a unique corpus by assembling the Portuguese corpora available.

References

1. Abdelali, A., Cowie, J., Soliman, H.: Building a modern standard Arabic corpus. In: Workshop on Computational Modeling of Lexical Acquisition, pp. 25–28 (2005)

2. do Amaral, D.O.F., Vieira, R.: NERP-CRF: uma ferramenta para o reconhecimento de entidades nomeadas por meio de conditional random fields. Linguamática **6**(1), 41–49 (2014)
3. Amri, S., Zenkouar, L., Outahajala, M.: Build a morphosyntaxically annotated amazigh corpus. In: Proceedings of the 2nd International Conference on Big Data, Cloud and Applications, p. 8. ACM (2017)
4. Cavalin, P., et al.: Building a question-answering corpus using social media and news articles. In: Silva, J., Ribeiro, R., Quaresma, P., Adami, A., Branco, A. (eds.) PROPOR 2016. LNCS (LNAI), vol. 9727, pp. 353–358. Springer, Cham (2016). https://doi.org/10.1007/978-3-319-41552-9_36
5. da Consolação Dias, C.: A análise de domínio, as comunidadesdiscursivas, a garantia de literatura e outras garantias. Informação Sociedade **25**(2) (2015)
6. Holthaus, P., et al.: How to address smart homes with a social robot? A multimodal corpus of user interactions with an intelligent environment. In: Proceedings of the Tenth International Conference on Language Resources and Evaluation (LREC 2016), pp. 3440–3446 (2016)
7. IBGE: Acesso à internet e à televisão e posse de telefone móvel celular para uso pessoal (2017). https://biblioteca.ibge.gov.br/index2.php/biblioteca-catalogo?view=detalhes&id=2101631
8. Mooers, C.N.: The next twenty years in information retrieval; some goals and predictions. Am. Doc. **11**(3), 229–236 (1960)
9. Mosley, M., Brackett, M.H., Earley, S., Henderson, D.: DAMA Guide to the Data Management Body of Knowledge. Technics Publications (2010)
10. Nelson, D.M.Q.: Uso de redes neurais recorrentes para previsão de séries temporais financeiras (2017)
11. Olah, C.: Understanding LSTM Networks (2015)
12. de Oliveira, M.G., de Souza Baptista, C., Campelo, C.E., Bertolotto, M.: A gold-standard social media corpus for urban issues. In: Proceedings of the Symposium on Applied Computing, pp. 1011–1016. ACM (2017)
13. Luz de Araujo, P.H., de Campos, T.E., de Oliveira, R.R.R., Stauffer, M., Couto, S., Bermejo, P.: LeNER-Br: a dataset for named entity recognition in Brazilian legal text. In: Villavicencio, A., et al. (eds.) PROPOR 2018. LNCS (LNAI), vol. 11122, pp. 313–323. Springer, Cham (2018). https://doi.org/10.1007/978-3-319-99722-3_32. https://cic.unb.br/~teodecampos/LeNER-Br/
14. Pirovani, J., Oliveira, E.: Portuguese named entity recognition using conditional random fields and local grammars. In: Proceedings of the Eleventh International Conference on Language Resources and Evaluation (LREC 2018) (2018)
15. Pretto, J.R.: O estilo jornalístico. Estudos Linguísticos **38**(3), 481–491 (2009)
16. Santos, J., Consoli, B., Santos, C., Terra, J., Collonini, S., Vieira, R.: Assessing the impact of contextual embeddings for Portuguese named entity recognition, pp. 437–442, October 2019. https://doi.org/10.1109/BRACIS.2019.00083
17. Souza, F., Nogueira, R., Lotufo, R.: Portuguese named entity recognition using BERT-CRF. arXiv preprint arXiv:1909.10649 (2019)
18. Spoustová, J., Spousta, M., Pecina, P.: Building a Web Corpus of Czech (2010)
19. Villalva, A., Mateus, M.H.M.: Morfologia do português. Universidade Aberta Lisboa (2008)
20. Zhou, P., et al.: Attention-based bidirectional long short-term memory networks for relation classification. In: Proceedings of the 54th Annual Meeting of the Association for Computational Linguistics (Volume 2: Short Papers), pp. 207–212 (2016)

Organizational Learning in the Age of Data

Andrew D. Banasiewicz[✉] ![ORCID]

Cambridge College, Boston, MA 02129, USA
`andrew.banasiewicz@cambridgecollege.edu`

Abstract. Traditionally, organizational learning has been viewed as a human-centric endeavor, but the rise of big data and advanced data analytic technologies are compelling a fundamental reconceptualization of the scope and modalities of organizational learning. Building on the foundation of explicit differentiation between episodic vs. ongoing learning inputs and new vs. cumulative learning outcomes, this article proposes a new typology of organizational learning modalities, which explicitly distinguishes between human reason-centric theoretical and experiential learning, and technology-centric computational and simulational learning modalities.

Keywords: Organizational learning · Computational learning · Simulational learning

1 Introduction

The widely used conception of organizational learning frames it as the process of acquiring, creating, integrating and distributing of information (Wang and Ellinger 2011; Dixon 1992; Huber 1991). Clearly visible in that characterization is that the ultimate purpose of organizational learning is to enhance the informational efficacy of managerial decision-making (Mattox 2016; Alegre et al. 2014; Greve 2003); somewhat less obvious is the implicit assumption that learning is an inherently human endeavor (Savory and Butterfield 1999; Ranyard et al. 1997). The latter, however, is gradually being called into question in view of the proliferation of self-learning algorithmic decision engines, commonly known as artificial intelligence (AI). Widely used to automate numerous routine business decisions, such as pricing of automotive insurance policies, eligibility approval for social or health services, or online product recommendations (Wright and Schultz 2018; Wichert 2014; Shi 2011), AI decision engines are capable of independently learning from data in a manner that fits the established conception of organizational learning. And given the already substantial and still rapidly expanding role and value of those technologies to organizational functioning and competitiveness, it is important to expressly include technology-based knowledge creation in the definition of organizational learning.

Broadly characterized, AI can be seen as a manifestation of progressively better attempts at mimicking the functioning of human brain (Agrawal 2014; Meyer et al. 2014), which suggests a number of similarities between human and machine learning, and some possible differences. For example, AI systems can perform neural network-like

R. Mugnaini (Ed.): DIONE 2020, LNICST 319, pp. 64–78, 2020.
https://doi.org/10.1007/978-3-030-50072-6_6

processes that emulate the innerworkings of the human brain, those systems are still not self-organizing or adaptive (Zhang 2010; Arbib 2003); at the same time, machine-based capture and retrieval of information far exceeds the capacity of human brain (Krishnaswamy and Sundarraj 2017; Stark and Tierney 2014). When considered from the standpoint of organizational learning, that mix of differences and similarities is suggestive of some important considerations, further accentuated by informational realities of the Age of Data. Most notably, significant synergies can be realized by leveraging the combined effect of the adaptiveness and creativity of human learning with the virtually unlimited information processing and retention, and the nearly instantaneous recall of any and all captured information that characterizes machine learning[1]. In fact, it is difficult to think of organizational competitiveness without considering organizational capabilities to capture, ingest, synthesize and disseminate decision-pertinent information to decision-responsible organizational stakeholders at the appropriate time.

1.1 The Rise and Fall of Organizational Learning

The recognition of learning as a distinct organizational competency can be traced back to the 1960s and 1970s (e.g., Arrow 1962; Cyert and March 1963; Cangelosi and Dill 1965; Argyris and Schon 1978), but it was not until the 1990s that the topics of organizational learning spurred wider interest among researchers (Rebelo and Gomes 2008; Bapuji and Crossan 2004). The resultant rich and varied research streams (e.g., Denton 1998; Popper and Lipshitz 1998, 2000; Cohen and Sproull 1996; Marquardt 1996; Dodgson 1993) included several popular books, such as the 1990 work titled *The Fifth Discipline: The Art and Practice of the Learning Organization,* which helped to popularize the notion of learning organization among practitioners. Unfortunately, the wider embrace of the idea of learning organizations has led to proliferation of 'tried and true' management solutions in the form of pre-packaged consulting frameworks chock-full of anecdotes, buzzwords, stories of great success, and other forms of management lore (Skeel 2005; Neuhauser 1998). Promising quick solutions to even intractable management problems, those templated conceptualizations, often accompanied by flawed axioms and other self-deceptive dictums became the de facto public face of organizational learning, effectively portraying an important organizational capability as yet another passing fad (Buckley et al. 2015; Robelo and Gomes 2008).

False prophets, buzzwords and sage anecdotes aside, to remain competitive in knowledge-driven economy firms have to develop and deploy robust means of creating and leveraging decision-guiding knowledge. When looked at from the standpoint of organizational survival, institutional learning capability can be seen as a manifestation of organizational adaptiveness (Thomas and Vohra 2015), enabling firms to fine-tune their behaviors (Templeton et al. 2009; Hult and Nichols 1996), and cognitive functioning capabilities (Chiva and Alegre 2005; Akgun et al. 2003). Within institutional setting, the behavioral and cognitive learning dimensions are shaped by a myriad of individual-level and system-wide influences, including organizational structure (Martínez-León

---

[1] The idea of technological singularity, or the merging of human and machine intelligence giving rise to infinitely more capable superintelligence, can be seen as much stronger expression of that hypothesis.

and Martínez-García 2011; Schreyogg and Sydow 2010; Hinnings et al. 1996; Gurpinar 2016), culture (Yates and de Oliveira 2016; Briley et al. 2014; Markus and Kitayama 1991) and group dynamics (Lucas and Kline 2008; Schein 1993), as well as an array of latent psychological and emotional characteristics (Lucas and Kline 2008; Schein 1993; Wastell 1999; Yanow 2000), and even biological traits (Salvador and Sadri 2018).

Equally important to developing sound organizational learning mechanism are the volume and variety of what is to be learned, as well as the available learning modalities. The rise of Big Data and the proliferation of large-scale data analytic capabilities produce torrents of new information, created on ongoing basis. Consequently, establishing of valid and reliable means of assessing and synthesizing ceaseless flows of large volumes of information constitutes an important aspect of organizational learning. In a very pragmatic sense, the ability to separate the proverbial chaff from grain, which amounts to finding and institutionalizing decision-guiding insights often hidden in masses of comparatively trivial informational tidbits, is an important prerequisite of effective learning. Another aspect of the modern data-rich and technology-enabled informational infrastructure is that, within organizational setting, the very manner in which learning takes place is expanding, now not only encompassing the traditionally human-centric modality, but also the rapidly maturing artificial, or technology-based learning capabilities.

2 Organizational Learning in the Age of Data

Implied in the traditional conception of learning is that it is a human-centric process, and since organizations are essentially human collectives, at its core, organizational learning is also commonly viewed as a human-centric undertaking (Marcus and Shoham 2014; Rasmussen and Nielsen 2011; Dixon 1992; Huber 1991). However, the rise of big data along with progressively more advanced and automated data analytic technologies are beginning to cast doubt on the validity of the human-centric conception of organizational learning, with machine learning (ML) and artificial intelligence (AI) as two obvious manifestations of the already substantial encroachment of non-human learning modalities.

Reframing of the notion of organizational learning is important not only from the standpoint of the somewhat obtuse concept validity (Neumayer and Plümper 2017; Locke 2012), but also because development of sound decision-making competencies is contingent on identification and utilization of all available and pertinent information (Dezi et al. 2018; Jeble et al. 2018; Cox et al. 2017; Laux et al. 2017; Eva 2015). Moreover, already considerable and growing share of organizational know-how resides outside of the human intellectual domain (Bolisani et al. 2018; Choi 2018), as illustrated by online recommendation engines and similar technologies. It thus follows that reconceptualization of the idea of organizational learning should start with clear and explicit delineation of learning inputs and outcomes.

2.1 Learning Inputs and Outcomes

In the most rudimentary sense, learning can be seen as a process the process of consuming inputs, in the form of various stimuli, with the goal of generating outputs, in the form of knowledge (Wang and Ellinger 2011). Broadly defined, learning process inputs can be either episodic, taking the form of ad hoc stimuli, or ongoing, manifesting themselves as recurring stimuli. Learning process outcomes, on the other hand, can take the form of incremental knowledge, or updates to existing knowledge. Consider Fig. 1.

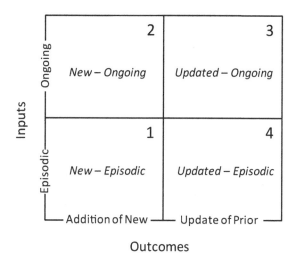

Fig. 1. Learning inputs and outcomes

The resultant 2 × 2 organizational learning input-outcome classification yields four distinct learning scenarios: new knowledge produced episodically (quadrant 1), new knowledge produced on ongoing basis (quadrant 2), ongoing update of prior knowledge (quadrant 3), and episodic update of prior knowledge (quadrant 4). Jointly, those four dimensions of organizational learning capture the 'what' aspect of learning in the form of distinct types of knowledge assets derived from different informational sources and learning modalities, such as theoretical understanding of a new phenomenon of interest derived from the most recent empirical research findings (quadrant 1) or the most recent data derived frequency of a particular type of insurance claims (quadrant 3). Essential to properly framing and contextualizing those distinct types of knowledge assets is a more in-depth detailing of the 'how' aspect of organizational learning, with particular emphasis on learning modalities.

2.2 Learning Modalities

In the most general sense, the basic tenets of human learning suggest that the ability to reason, conceptualized as the power of the mind to think and understand by a process of logic (Kahneman 2011; Contreras 2010) is most emblematic of human learning, while the ability to identify patterns in vast quantities of data is most descriptive of machine learning (Wright and Schultz 2018; Wichert 2014; Shi 2011). That basic distinction is captured in two meta-categories of organizational learning: 'reason-based', which embodies individual and collective cognitive and behavioral knowledge, and 'technology-based', which embodies the efficacy of the broadly defined informational infrastructure to autonomously perform specific, typically routine, organizational tasks. Each of the two meta-categories can be further subdivided into more operationally meaningful categories of tacit and explicit, and computational and simulational learning, for reason- and technology-based, respectively. Figure 2 below offers a graphical representation of the resultant typology of organizational learning.

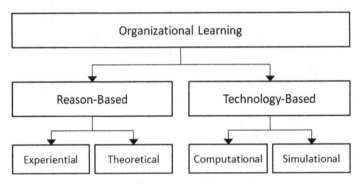

Fig. 2. The new typology of organizational learning

Although outlined in Fig. 2 as four distinct manifestations of organizational learning, experiential, theoretical, computational and simulational learning modes can also be thought of as progressively more sophisticated means of knowledge creation. From early man gazing at the stars and trying to make sense of natural phenomena (experiential learning), to early philosophers and scientists discerning the underlying laws of nature (theoretical learning), to modern data- and technology-enabled investigators identifying new and testing presumed relationships (computational learning), to the now-emerging means of simulating reality as a mean of learning that transcends experience and physical reality (simulational learning). And although experiential, theoretical, computational and simulational learning can be seen progressive more sophisticated means of learning, it is important to think of them as complements, not replacements, in the same way that planes, automobiles, bicycles and simply walking are all different means of traversing distance.

Reason-Based Learning. At an individual level, learning can be broadly characterized as acquisition of new or reinforcement existing knowledge (Giorgis and Johnson 2001; Estes 1956). The underlying process begins with awareness-arousing stimulus being encoded into short-term memory in one of two forms: iconic, or visual, and echoic, or auditory. The process of learning is then initiated: It starts with the formation of new neuronal connections, which is followed by consolidation, or strengthening and storing of remembrances as long-term memories, with distinct clusters of neurons being responsible for holding different types of knowledge (Ehrlich and Josselyn 2016; Zull 2002). Any subsequent retrieval of memories from long-term to active memory brings about re-consolidation, or strengthening of the stored knowledge, often referred to as remembering (Van Dam 2013; Leamnson 2000).

In a more abstract sense, learning can be characterized as adding new or modifying information already stored in memory based on new input or experiences (Kesner and Martinez 2007; Jean-Marie and Tidball 2006). It is an active process involving sensory input to the brain coupled with extraction of meaning from sensory input; it is also a fluid process, in the sense each subsequent experience prompts the brain to (subconsciously) reorganize stored information, effectively reconstituting its contents through a repetitive updating procedure known as brain plasticity (Brynie 2009; Moller 2009; Kolb and Whishaw 1998). Though generally resulting in improvements to existing knowledge, brain plasticity can nonetheless bring about undesirable outcomes, most notably in the form continuous re-casting of memories – in fact, that is one of the reasons eyewitness accounts become less and less reliable with the passage of time. All considered, the widely used characterization of learning as the 'acquisition of knowledge' oversimplifies what actually happens when new information is added into the existing informational mix – rather than being simply 'filed away' and stored in isolation, any newly acquired information is instead integrated into a complex web of existing knowledge.

Within the confines of reason-based learning, the most elementary knowledge acquisition mechanism entails immersion in, or observation of, a process or a phenomenon (Brito and Barros 2005; Merav 1999), commonly referred to as experiential learning. Subjective and situational, this mode of learning can be seen as a product of a curious mind driven to understand the nature of a particular experience, and it is built around systematic examination of sensory experiences, particularly those obtained by means of direct observation or hands-on participation (Douglas Greer et al. 2006; Braaksma et al. 2002; Blandin et al. 1999). Being entirely shaped by person-specific factors, experiential learning implicitly dismisses existence of innate – i.e., generalizable – ideas, resulting in knowledge that is entirely defined by individual learners (Gascoigne and Thornton 2014; Rainbird et al. 2004). That mode of learning is particularly important in the context of specific tasks, such as underwriting of executive risk or managing retail customer loyalty programs.

Complementing the experiential learning dimension of reason-based learning is theoretical learning, which is focused primarily on common knowledge, or innate ideas that transcend individual experiences. It entails developing an understanding of universally true and commonly accepted abstract formulations and explanations, as exemplified by the axioms and rules of mathematics or the laws of nature (Brante et al. 2015; Karpov and Bransford 1995). That mode of learning typically plays a very important role in the attainment of professional competence, as evidence by numerous professional certification requirements (Blank et al. 2012; Sense 2008).

Plasticity, Bias and Channel Capacity. In a very general sense, knowledge can be thought of as a library – a collection of systematic, procedural and episodic remembrances acquired via explicit and tacit learning (Kahin and Foray 2006). However, as suggested by the notion of brain plasticity, unlike physical libraries, neural networks-stored 'collections' are subject to ongoing re-shaping, triggered by the process of assimilating of new memories. The resultant continuous re-writing of old memories means that an individual-level effective topical knowledge is ever-changing (which is why eye witness accounts become less and less reliable with the passage of time), and that the ongoing interpretation and re-interpretation of knowledge can exert a profound impact on individuals' perception and judgment.

While the ongoing re-shaping of knowledge affects the validity and reliability of individuals' knowledge, cognitive bias impacts the manner in which stored information is used (Caputo 2016; Hilbert 2012; Kahneman 2011). Reasoning distortions such as availability heuristic (a tendency to overestimate the importance of available information) or confirmation bias (favoring of information that confirms one's pre-existing beliefs) attest to the many ways subconscious information processing mechanics can warp the manner in which overtly objective information shapes individual-level sense-making. To make matters worse, unlike machines that 'remember' all information stored in them equally well at all times, the brain's persistent self-rewiring renders older, not sufficiently reinforced memories progressively 'fuzzier' and more difficult to retrieve (Brynie 2009; Moller 2009). As a result, human recall tends to be incomplete and selective.

Moreover, the amount of information human brain can cognitively process in attention at any given time is limited due to a phenomenon known as human channel capacity (Benish 2015; Woungang et al. 2010). Research suggests that, on average, a person can actively consider approximately 7 ± 2 of discrete pieces of information (Massa and Keston 1965; Miller 1956). When coupled with the ongoing reshaping of previous learnings (brain plasticity) and the possibly distorted nature of perception (cognitive bias), channel capacity brings to light cognitively-biological human reasoning limitations.

Emotion and Motivation. Looking beyond factors that capture some of the brain mechanics related reasoning limitations outlined earlier, reason-based learning is also impacted by numerous attitudinal factors, most notably related to emotions and motivation (Kahneman 2011; Sessa and London 2008). For instance, more positive experiences tend to manifest themselves in more complete recollections than negative events, and those events that occurred more recently appear to be more significant or thus likely to recur. Moreover, desire to perform better has been shown to lead to deeper learning, even when time spent on learning, as well as learners' gender and ability were controlled for, highlighting the importance of intrinsic motivation to learning (Everaert et al. 2017). While commonly considered in the context of individual-level characteristics, emotion and motivation also have important group-level analogs, outlined next.

Group Dynamics. Contradicting conventional wisdom which suggests that groups make better decisions than individuals, research in areas of social cognition and social psychology instead suggests that hat groups do not always outperform individual, and that a combination of cognitive, social and situational influences are ultimately stronger determinants of the quality of decision-making (Cristofaro 2017; Mazutis and Eckardt 2017; Bhatt 2000). Higher levels of confidence often associated with group decisions may not yield higher decision quality because of a phenomenon known as 'groupthink', or a dysfunctional pattern of thought and interaction characterized by closed-mindedness and uniformity expectations (Russell et al. 2015; Schafer and Crichlow 1996), and biased information search, characterized by strong preference for information that supports the group's view (Kopsacheilis 2018; Rozas 2012; Fischer et al. 2011).

A yet another important, organizational decision-making related aspect of group dynamics is group conflict (Katz et al. 2016; Stanley 1981). As suggested by social exchange theory, which views the stability of group interactions through a theoretical lens of negotiated exchange between parties, individual group members are ultimately driven by the desire to maximize their benefits, thus conflict tends to arise when group dynamics take on more competitive than collaborative character (Li-Fen 2008; Gould-Williams and Davies 2005). Keeping in mind that the realization of group decision-making potential requires full contributory participation on the part of individual group members, within-group competition reduces the willingness of individuals to contribute their best to the group effort. Not only can that activate individuals' fears of being exploited, as well as heighten the desire to exploit others, it can compel individuals to become more focused on standing out in comparison with others. That can activate tendencies to evaluate one's own information more favorably than that others' (Arai et al. 2016; Van Swol 2007), and also to evaluate more positively any information that is consistent with one's initial preferences (Faulmüller et al. 2010; Mojzisch and Schulz-Hardt 2010).

Technology-Based Learning. The growing sophistication and proliferation of self-learning technologies, commonly referred to as artificial intelligence (AI), is beginning to challenge the traditional, human-centric conception of organizational learning (Lowe and Sandamirskaya 2018; Betzler 2016; Estes 1956). Machine learning, a sub-category of AI that focuses on endowing computers with the ability to learn without being expressly programmed, discern patterns from available data, accumulate and synthesize the resultant knowledge, and then execute specific tasks using self-discerned decision logic (Shandilya 2014; Usuelli 2014; Witten et al. 2011). In fact, as implied in the term 'artificial intelligence', AI systems are expressly designed to mimic the functioning of the human brain, as illustrated by one of the more commonly used artificial learning approaches represented by a family of algorithms known as neural networks. Unimpeded by human limitations in the form of cognitive bias, fatigue or channel capacity, and taking advantage of practically limitless computational resources, AI is pushing the broadly defined ability to learn beyond the traditional limitations of human-centric information processing (Krishnaswamy and Sundarraj 2017; Stark and Tierney 2014). And in some context, most notably when performing routine, repetitive tasks, AI-based decision engines can in fact outperform humans, primarily because those systems can rapidly, tirelessly and objectively infer from the often vast quantities of data decision alternatives that exhibit the highest probability of desired outcomes (Usuelli 2014; Witten et al. 2011).

It is important to emphasize that technology-based learning is a complement, not a replacement for human learning. When decisions are characterized as repetitive and structured, and the decision-making environment is characterized as stable, technology-based learning can offer incremental value to the organization by enabling more exhaustive and objective utilization of the available data. Conversely, there are many decision situations in which technology-based learning offers considerably less beneficial, as is the case when available historical data have limited predictive value, and/or decision environment is highly volatile. That said, organizations typically face a mix of repetitive-structured-stable and ad hoc decisions, suggesting that technology-based learning should be considered an important aspect of the overall organizational learning strategy.

Also comprised of two complementing dimensions, technology-based learning can take the form of computational (Andreopoulos and Tsotsos 2013; Suykens 2003) or simulational (Hey et al. 2009; Wood et al. 2009) learning. While overtly quite similar in the sense that both learning modalities are built on the foundation of analysis of raw data using sophisticated data analytic tools and techniques, computational learning is primarily focused on the 'what-is' dimension of knowledge creation, while simulational learning explores the more speculative 'what-if' dimension of data analytic knowledge. More concretely, the former takes the form of informational summarization and pattern identification, while the latter is built around anticipatory, forward-looking data-based simulations of future outcomes of interest. Utility-wise, computational learning is invaluable to guiding recurring, routine decisions characterized by high degrees of longitudinal stability, as exemplified by managing insurance claims, while simulational learning is essential to infusing objectivity into non-routine decisions, as exemplified by emergence of disruptive technologies.

Simulational learning can be thought of as machine equivalent of human reason-based theoretical learning. More specifically, it enables constructed reality-based knowledge creation, or discovery of universal generalizations within artificial representations of the world, broadly referred to as virtual reality, perhaps best exemplified by astrophysical research delving into the birth of our physical universe. Virtual reality-enabled learning makes possible generation of previously inaccessible insights (e.g., conditions that existed shortly after the Big Bang) because it enables the creation of possible but not-yet-observed situations, and virtually limitless what-if type of scenario planning.

Overabundance. In the most rudimentary sense, data can be conceptualized as a mix of signal, which is potentially informative, and noise, which is generally non-informative (Subedi 2013; Woodward 2010). Hence one of the core aspects of data utilization is to separate signal from noise, a task that becomes increasingly more challenging as the volume and variety of available data expand (Jain and Sharma 2014; Sinha 2014). Walmart, the world's largest retailer, handles more than a million customer transactions per hour; by 2020, the aggregate volume of business-to-business and business-to-consumer transactions is expected to surpass 450 billion per day (Nadkarni and Mehra 2018). And many of those transacting consumer, more than 5 billion as of 2018, are calling, texting, tweeting and browsing on mobile devices, all of which adds informationally-rich pre- and post-purchase details (Fenwick and Schadler 2018). However, given that the bulk of data available to organizations represents is a product of passive recording of an ever-growing array of states and events (Wiggins 2012; Tupper 2011), finding the few organizational decision-related insights typically entails analytically sifting through vast quantities of non-informative noise.

The often staggeringly large quantities of available data are perhaps the most visible manifestation of the difficulty of separating information from noise. However, within the confines of technology-based learning, epistemology, or the essence of validity and reliability of what is considered 'knowledge', poses an even more formidable challenge. Lacking the face validity or 'does it make sense' aspect of the reason-based learning, technology-based learning has to rely on generalizable decision heuristics to enable the automated algorithms to independently and consistently differentiate between material and spurious conclusions. Consider is common scenario: A computer algorithm sifting through data in search for material patterns pinpoints a recurring association between X and Y – once identified, the association is 'learned' and subsequently used as a driver of the algorithm-enabled decision-making. However, there is often a non-trivial possibility that what manifested itself as a recurring association between X and Y is erroneous, due to both X and Y being influenced by unaccounted for (i.e., not captured in the available data) factor Z, effectively rendering the presumed association illusory (Szatkowski and Rosiak 2014). Moreover, even if the X-Y association is unaffected by the unaccounted for factor Z, the widely used statistical significance testing may produce falsely positive conclusions. More specifically, the often large number of records used in analyses can result in magnitudinally trivial effect size, such as a correlation coefficient, being deemed material, or statistically significant, because of the well-known dependence of statistical significance tests on sample size (Banasiewicz 2013).

3 Conclusions and Recommendations

In information-driven economy, few organizational competencies are as important as the capability to systematically capture, synthesize and disseminate throughout the organization competitively advantageous decision-guiding knowledge. Historically viewed as a human-centric endeavor, organizational learning is being re-defined by the rise of big data and advanced data analytic technologies, all of which is compelling a fundamental reconceptualization of the scope and modalities of organizational learning, and research summarized here offers a revised and expanded conceptualization of that important organizational competency. Building on the foundation of explicit differentiation between episodic vs. ongoing learning inputs and new vs. cumulative learning outcomes, a new typology of organizational learning modalities is proposed, which explicitly distinguishes between human reason-centric theoretical and experiential learning, and technology-centric computational and simulational learning modalities. By expressly encompassing artificial intelligence, machine learning and other manifestations of technology-based learning, the proposed organizational learning typology offers a more comprehensive and timely framing of organizational learning. By acknowledging the distinctiveness of human reason- and technology-based learning modalities, business organizations will be able to develop more robust and effective systems and mechanisms to support their goal of remaining competitive in knowledge-based economy.

As is the case with all conceptual frameworks, the conceptualization offered here needs to be subjected to theoretical and application scrutiny. Does the new typology of organizational learning summarized in Fig. 2 exhibit the necessary MECE (mutually exclusive and collective exhaustive) characteristics? Is the said typology practically meaningful, or stated differently, does the use of the new typology of organizational learning lead to measurable gains in organizational learning efficacy? Those and related questions need to be answered to ascertain both the theoretical and practical values of the proposed framework.

References

Agrawal, D.: Analytics based decision making. J. Indian Bus. Res. **6**(4), 332–340 (2014)
Akgun, A.E., Lynn, G.S., Byrne, J.C.: Organizational learning: a socio-cognitive framework. Hum. Relat. **56**(7), 839–868 (2003)

Alegre, J., Chiva, R., Fernández-Mesa, A., Ferreras-Méndez, J.: Shedding New Lights on Organisational Learning, Knowledge and Capabilities. Cambridge Scholars Publishing, Newcastle upon Tyne (2014)

Alpar, P., Schulz, M.: Self-service business intelligence. Bus. Inf. Syst. Eng. **58**(2), 151–155 (2016)

Andreopoulos, A., Tsotsos, J.: A computational learning theory of active object recognition under uncertainty. Int. J. Comput. Vis. **101**(1), 95–142 (2013)

Arai, A., Fan, Z., Dunstan, M., Shibasaki, R.: Comparative perspective of human behavior patterns to uncover ownership bias among mobile phone users. Int. J. Geo-Inform. **5**(6), 1–12 (2016)

Arbib, M.: The Handbook of Brain Theory and Neural Networks, 2nd edn. MIT Press, Cambridge (2003)

Argyris, C., Schon, D.A.: Organizational Learning: A Theory of Action Perspective. Addison-Wesley, Reading (1978)

Arora, A.: The "organization" as an interdisciplinary learning zone: using a strategic game to integrate learning about supply chain management and advertising. Learn. Organ. **19**(2), 121–133 (2012)

Arrow, K.: The economic implications of learning by doing. Rev. Econ. Stud. **29**(3), 155–173 (1962)

Banasiewicz, A.D.: Marketing Database Analytics: Transforming Data for Competitive Advantage. Routledge, New York (2013)

Bapuji, H., Crossan, M.: From questions to answers: Reviewing organizational learning research. Manag. Learn. **35**(4), 397–417 (2004)

Benish, W.: The channel capacity of a diagnostic test as a function of test sensitivity and test specificity. Stat. Methods Med. Res. **24**(6), 1044–1052 (2015)

Benson, J., Dresdow, S.: Systemic decision application: linking learning outcome assessment to organizational learning. J. Workplace Learn. **10**(6/7), 301–307 (1998)

Betzler, R.: Is statistical learning a mechanism? Philos. Psychol. **29**(6), 826–843 (2016)

Bhatt, G.D.: Information dynamics, learning and knowledge creation in organizations. Learn. Organiz. **7**(2), 89–98 (2000)

Blandin, Y., Lhuisset, L., Proteau, L.: Cognitive processes underlying observational learning of motor skills. Q. J. Exp. Psychol. **52**(4), 957–979 (1999)

Blank, J., Van Hulst, B., Koot, P., Van der Aa, R.: Benchmarking overhead in education: a theoretical and empirical approach. Benchmarking Int. J. **19**(2), 239–254 (2012)

Bolisani, E., Scarso, E., Padova, A.: Cognitive overload in organizational knowledge management: case study research. Knowl. Process Manag. **25**(4), 223–231 (2018)

Braaksma, M., Rijlaarsdam, G., Van den Bergh, H.: Observational learning and the effects of model-observer similarity. J. Educ. Psychol. **94**(2), 405–415 (2002)

Brante, G., Holmqvist Olander, M., Holmquist, P., Palla, M.: Theorising teaching and learning: pre-service teachers' theoretical awareness of learning. Eur. J. Teach. Educ. **38**(1), 102–118 (2015)

Briley, D.A., Wyer, R.S., Li, E.: A dynamic view of cultural influence: a review. J. Consum. Psychol. **24**(4), 557–571 (2014)

Brito, P., Barros, C.: Learning-by-consuming and the dynamics of the demand and prices of cultural goods. J. Cult. Econ. **29**(2), 83–106 (2005)

Brynie, F.: Brain Sense: The Science of the Senses and How We Process the World Around Us. American Management Association, New York (2009)

Buckley, M., et al.: Management lore continues alive and well in the organizational sciences. J. Manag. Hist. **21**(1), 68–97 (2015)

Campbell, D.T., Stanley, J.C.: Experimental and Quasi-experimental Designs for Research. Rand McNally, Skokie (1966)

Cangelosi, V.E., Dill, W.R.: Organizational learning: observations toward a theory. Adm. Sci. Q. **10**(2), 175–203 (1965)

Caputo, A.: Overcoming judgmental biases in negotiations: a scenario-based survey analysis on third party direct intervention. J. Bus. Res. **69**(10), 4304–4312 (2016)

Cheung-Blunden, V., Khan, S.: A modified peer rating system to recognise rating skill as a learning outcome. Assess. Eval. High. Educ. **43**(1), 58–67 (2018)

Chiva, R., Alegre, J.: Organizational learning and organizational knowledge: towards the integration of two approaches. Manag. Learn. **36**(1), 49–68 (2005)

Choi, S.: Organizational knowledge and information technology: the key resources for improving customer service in call centers. Inf. Syst. E-Bus. Manag. **16**(1), 187–203 (2018)

Cohen, M.D., Sproull, L.E.: Organizational Learning. Sage, Thousand Oaks (1996)

Contreras, D.: Psychology of Thinking. Nova Science, New York (2010)

Cox, B., et al.: Lip service or actionable insights? Linking student experiences to institutional assessment and data-driven decision making in higher education. J. High. Educ. **88**(6), 835–862 (2017)

Cristofaro, M.: Reducing biases of decision making processes in complex organizations. Manag. Res. Rev. **40**(3), 270–291 (2017)

Cyert, R.M., March, J.G.: A Behavioral Theory of the Firm. Prentice-Hall, Englewood Cliffs (1963)

Denton, J.: Organizational Learning and Effectiveness. Routledge, London (1998)

Dezi, L., Santoro, G., Gabteni, H., Pellicelli, A.: The role of big data in shaping ambidextrous business process management. Bus. Process Manag. J. **24**(5), 1163–1175 (2018)

Dixon, N.M.: Organizational learning: a review of literature with implications for HRD professionals. Hum. Resour. Dev. Q. **3**(1), 29–49 (1992)

Dodgson, M.: Organizational learning: a review of some literatures. Organ. Stud. **14**(3), 375–394 (1993)

Douglas Greer, R., Dudek-Singer, J., Gautreaux, G.: Observational learning. Int. J. Psychol. **41**(6), 486–499 (2006)

Dubey, R., Gunasekaran, A., Childe, S., Papadopoulos, T.: Skills needed in supply chain-human agency and social capital analysis in third party logistics. Manag. Decis. **56**(1), 143–159 (2018)

Ehrlich, D., Josselyn, S.: Plasticity-related genes in brain development and amygdala-dependent learning. Genes Brain Behav. **15**(1), 125–143 (2016)

Estes, W.: Learning. Annu. Rev. Psychol. **7**(1), 1–38 (1956)

Eva, G.: Framing skilful performance to enact organizational knowledge: Integrating data-driven and user-driven practice. Management **10**(3), 255–271 (2015)

Everaert, P., Opdecam, E., Maussen, S.: The relationship between motivation, learning approaches, academic performance and time spent. Acc. Educ. **26**(1), 78–107 (2017)

Faulmüller, N., Kerschreiter, R., Mojzisch, A., Schulz-Hardt, S.: Beyond group-level explanations for the failure of groups to solve hidden profiles: the individual preference effect revisited. Group Process. Intergroup Relat. **13**(5), 653–671 (2010)

Fenwick, N., Schadler, T.: Digital rewrites the rules of business. The Vision Report in the Digital Business Playbook, Forrester, 26 February 2018. https://www.forrester.com/report/Digital+Rewrites+The+Rules+Of+Business/-/E-RES137090. Accessed 30 Nov 2018

Fischer, P., Lea, S., Kastenmuller, A., Greitemeyer, T., Fischer, J., Frey, D.: The process of selective exposure: why confirmatory information search weakens over time. Organ. Behav. Hum. Decis. Process. **114**(1), 37–48 (2011)

Gascoigne, N., Thornton, T.: Tacit Knowledge. Routledge, London (2014)

Giorgis, C., Johnson, N.: The learning process. Read. Teach. **55**(1), 86–94 (2001)

Gould-Williams, J., Davies, F.: Using social exchange theory to predict the effects of HRM practice on employee outcomes. Public Manag. Rev. **7**(1), 1–24 (2005)

Greve, H.: Organizational Learning from Performance Feedback: A Behavioral Perspective on Innovation and Change. Cambridge University Press, Cambridge (2003)

Gurpinar, E.: Organizational forms in the knowledge economy: a comparative institutional analysis. J. Evol. Econ. **26**(1), 501–518 (2016)

Hart, R., Hiltbrand, T.: Bridging the analytics skill gap with crowdsourcing. Bus. Intell. J. **19**(2), 8–15 (2014)

Hey, T., Tansley, S., Tolle, K. (eds.): The Fourth Paradigm: Data-Intensive Scientific Discovery. Microsoft Research, Redmond (2009)

Hilbert, M.: Toward a synthesis of cognitive biases: how noisy information processing can bias human decision making. Psychol. Bull. **138**(2), 211–237 (2012)

Hinnings, C.R., Thibault, L., Slack, T., Kikulis, L.M.: Values and organizational structure. Hum. Relat. **49**(7), 885–916 (1996)

Huber, G.P.: Organizational learning: the contributing process and the literatures. Organ. Sci. **2**(1), 88–115 (1991)

Hult, G.T.M., Nichols, E.L.: The organizational buyer behavior learning organization. Ind. Mark. Manage. **25**(3), 197–207 (1996)

Jain, P., Sharma, P.: Behind Every Good Decision: How Anyone Can Use Business Analytics to Turn Data Into Profitable Insight. (Edited by L. Jayaraman). American Management Association, New York (2014)

Jean-Marie, A., Tidball, M.: Adapting behaviors through a learning process. J. Econ. Behav. Organ. **60**(3), 399–422 (2006)

Jeble, S., Dubey, R., Childe, S., Papadopoulos, T., Roubaud, D., Prakash, A.: Impact of big data and predictive analytics capability on supply chain sustainability. Int. J. Logist. Manag. **29**(2), 513–538 (2018)

Kahin, B., Foray, D.: Advancing Knowledge and the Knowledge Economy. MIT Press, Cambridge (2006)

Kahneman, D.: Thinking, Fast and Slow. Farrar, Straus and Giroux, New York (2011)

Karpov, Y., Bransford, J.: L.S. Vygotsky and the doctrine of empirical and theoretical learning. Educ. Psychol. **30**(2), 61–66 (1995)

Katz, N.H., Sosa, K.J., Harriott, S.A.: Overt and covert group dynamics: an innovative approach for conflict resolution preparation. Conflict Resolut. Q. **33**(3), 313–348 (2016)

Kesner, R., Martinez, J.: Neurobiology of Learning and Memory, 2nd edn. Academic Press, Amsterdam (2007)

Kolb, B., Whishaw, I.: Brain plasticity and behavior. Annu. Rev. Psychol. **49**(1), 43–64 (1998)

Kopsacheilis, O.: The role of information search and its influence on risk preferences. Theory Decis. Int. J. Multidiscip. Adv. Decis. Sci. **84**(3), 311–339 (2018)

Krishnaswamy, V., Sundarraj, R.: Organizational implications of a comprehensive approach for cloud-storage sourcing. Inf. Syst. Front. J. Res. Innov. **19**(1), 57–73 (2017)

Laux, C., Li, N., Seliger, C., Springer, J.: Impacting big data analytics in higher education through six sigma techniques. Int. J. Prod. Perform. Manag. **66**(5), 662–679 (2017)

Leamnson, R.: Learning as biological brain change. Change **32**(6), 34–40 (2000)

Li-Fen, L.: Knowledge-sharing in R&D departments: a social power and social theory perspective. Int. J. Hum. Resour. Manag. **19**(10), 1881–1895 (2008)

Locke, E.: Construct validity vs. concept validity. Hum. Resour. Manag. Rev. **22**(2), 146–148 (2012)

Lowe, R., Sandamirskaya, Y.: Learning and adaptation: neural and behavioural mechanisms behind behaviour change. Connect. Sci. **30**(1), 1–4 (2018)

Lucas, C., Kline, T.: Understanding the influence of organizational culture and group dynamics on organizational change and learning. Learn. Organ. **15**(3), 277–287 (2008)

Marcus, T., Shoham, S.: Knowledge assimilation by employees in learning organizations. Learn. Organ. **21**(6), 350–368 (2014)

Markus, H.R., Kitayama, S.: Culture and the self: implications for cognition, emotion, and motivation. Psychol. Rev. **98**(2), 224–253 (1991)

Marquardt, M.: Building the Learning Organization. McGraw-Hill, New York (1996)

Martínez-León, M.I., Martínez-García, J.: The influence of organizational structure on organizational learning. Int. J. Manpower **32**(5–6), 537–566 (2011)

Massa, R., Keston, R.: Minimum attention display techniques. Navigation **12**(2), 153–163 (1965)

Mattox II, J.: Learning Analytics: Measurement Innovations to Support Employee Development. Kogan Page, Philadelphia (2016)

Mazutis, D., Eckardt, A.: Sleepwalking into catastrophe: cognitive biases and corporate climate change inertia. Calif. Manag. Rev. **59**(3), 74–108 (2017)

Merav, A.: Perceptual learning. Curr. Dir. Psychol. Sci. **8**(4), 124–128 (1999)

Meyer, G., et al.: A machine learning approach to improving dynamic decision making. Inf. Syst. Res. **25**(2), 239–263 (2014)

Miller, G.A.: The magical number seven plus or minus two: some limits on our capacity for processing information. Psychol. Rev. **101**(2), 343–352 (1956)

Mojzisch, A., Schulz-Hardt, S.: Knowing others' preferences degrades the quality of group decisions. J. Pers. Soc. Psychol. **98**(5), 794–808 (2010)

Moller, A.: The Malleable Brain: Benefits and Harm from Plasticity of the Brain. Nova Biomedical Books, New York (2009)

Nadkarni, A., Mehra, R.: IDC FutureScape: Worldwide enterprise infrastructure 2019 predictions. IDC Report, November 2018

Neuhauser, P.: Corporate Legends and Lore: The Power of Storytelling as a Management Tool. Peg G. Neuhauser, Austin (1998)

Neumayer, E., Plümper, T.: Robustness Tests for Quantitative Research. Cambridge University Press, Cambridge (2017)

Popper, M., Lipshitz, R.: Organizational learning mechanisms: a structural and cultural approach to organizational learning. J. Appl. Behav. Sci. **31**(2), 181–196 (1998)

Popper, M., Lipshitz, R.: Organizational learning: mechanisms, culture, and feasibility. Manag. Learn. **31**(2), 181–196 (2000)

Rainbird, H., Fuller, A., Munro, A.: Workplace Learning in Context. Routledge, London (2004)

Ranyard, R., Crozier, W., Svenson, O.: Decision Making: Cognitive Models and Explanations. Routledge, London (1997)

Rasmussen, P., Nielsen, P.: Knowledge management in the firm: Concepts and issues. Int. J. Manpower **32**(5/6), 479–493 (2011)

Rebelo, T.M., Gomes, A.D.: Organizational learning and the learning organization: reviewing evolution for prospecting the future. Learn. Organ. **15**(4), 294–308 (2008)

Rozas, J.: Biased information and effort. Econ. Inq. **50**(2), 484–501 (2012)

Russell, J., Hawthorne, J., Buchak, L.: Groupthink. Philos. Stud. **172**(5), 1287–1309 (2015)

Salvador, R., Sadri, G.: The biology of decision-making. Ind. Manag. Jan/Feb, 12–17 (2018)

Savory, A., Butterfield, J.: Holistic Management: A New Framework for Decision Making, 2nd edn. Island Press, Washington, D.C. (1999)

Schafer, M., Crichlow, S.: Antecedents of groupthink: a quantitative study. J. Conflict Resol. **40**(3), 415–435 (1996)

Schein, E.H.: On dialogue, culture, and organizational learning. Organ. Dyn. **22**, 40–51 (1993)

Schreyogg, G., Sydow, J.: Organizing for fluidity? Dilemmas of new organizational forms. Organ. Sci. **21**(6), 1251–1262 (2010)

Sense, A.: Conceptions of learning and managing the flow of knowledge in the project-based environment. Int. J. Manag. Proj. Bus. **1**(1), 33–48 (2008)

Sessa, V.I., London, M.: Interventions to stimulate group learning in organizations. J. Manag. Dev. **27**(6), 554–573 (2008)

Shandilya, S. (ed.): Advances in Machine Learning Research. Novinka, New York (2014)

Shi, Z.: Advanced Artificial Intelligence. World Scientific, Singapore (2011)

Sinha, S.: Making Big Data Work for Your Business: A Guide to Effective Big Data Analytics. Impackt Publishing, Birmingham (2014)

Skeel Jr., D.: Icarus in the Boardroom: The Fundamental Flaws in Corporate America and Where They Came From. Oxford University Press, Oxford (2005)

Subedi, D.: Signal and noise: why so many predictions fail – but some don't. Compet. Rev. Int. Bus. J. 23(4–5), 426–430 (2013)

Suykens, J.: Advances in Learning Theory: Methods, Models, and Applications. NATO Science Series, Series III: Computer and Systems Sciences, vol. 190. IOS Press, Amsterdam (2003)

Stanley, J.D.: Dissent in organizations. Acad. Manag. Rev. 6(1), 13–19 (1981)

Stark, L., Tierney, M.: Lockbox: mobility, privacy and values in cloud storage. Ethics Inf. Technol. 16(1), 1–13 (2014)

Szatkowski, M., Rosiak, M.: Substantiality and Causality. De Gruyter, Berlin (2014)

Templeton, G.F., Schmidt, M.B., Taylor, G.S.: Managing the diffusion of organizational learning behavior. Inf. Syst. Front. 11(2), 189–200 (2009)

Thomas, N., Vohra, N.: Three debates in organizational learning: what every manager should know. Dev. Learn. Organ. Int. J. 29(3), 3–6 (2015)

Tuckman, B.W.: Conducting Educational Research, 3rd edn. Harcourt Brace Jovanovich, San Diego (1988)

Tupper, C.: Data Architecture: From Zen to Reality. Morgan Kaufmann, Amsterdam (2011)

Usuelli, M.: R Machine Learning Essentials: Gain Quick Access to the Machine Learning Concepts and Practical Applications Using the R Development Environment. Packt Publishing, Birmingham (2014)

Van Dam, N.: Inside the learning brain. T + D 67(4), 30–35 (2013)

Van Swol, L.: Perceived importance of information: the effects of mentioning information, shared information bias, ownership bias, reiteration, and confirmation bias. Group Process. Intergroup Relat. 10(2), 239–256 (2007)

Wang, Y.L., Ellinger, A.D.: Organizational learning: perception of external environment and innovation performance. Int. J. Manpower 32(5/6), 512–536 (2011)

Wastell, D.G.: Learning dysfunctions in information systems development: overcoming the social defences with transitional objects. MIS Q. 23, 581–600 (1999)

Wichert, A.: Principles of Quantum Artificial Intelligence. World Scientific, Hackensack (2014)

Wiggins, B.: Effective Document and Data Management: Unlocking Corporate Content, 3rd edn. Gower, Farnham (2012)

Witten, I., Frank, E., Hall, M.: Data Mining: Practical Machine Learning Tools and Techniques, 3rd edn. Morgan Kaufmann, Burlington (2011)

Wood, R., Beckmann, J., Birney, D.: Simulations, learning and real world capabilities. Educ. Train. 51(5/6), 491–510 (2009)

Woodward, J.: Data, phenomena, signal, and noise. Philos. Sci. 77(5), 792–803 (2010)

Woungang, I., Misra, S., Misra, S.: Selected Topics in Information and Coding Theory. World Scientific, Singapore (2010)

Wright, S., Schultz, A.: The rising tide of artificial intelligence and business automation: developing an ethical framework. Bus. Horiz. 61(6), 823–832 (2018)

Yanow, D.: Seeing organizational learning: a "cultural" view. Organization 7, 247–268 (2000)

Yates, J.F., de Oliveira, S.: Culture and decision making. Organ. Behav. Hum. Decis. Process. 136, 106–118 (2016)

Zhang, W.: Computational Ecology: Artificial Neural Networks and Their Applications. World Scientific, Singapore (2010)

Zull, J.: The Art of Changing the Brain: Enriching Teaching by Exploring the Biology of Learning. Stylus Publishing, Sterling (2002)

Identification of the Relationships Between the Stages of the Data Lifecycle and the Principles of the Brazilian General Data Protection Act

Gislaine Parra Freund[(✉)] , Priscila Basto Fagundes ,
and Douglas Dyllon Jeronimo de Macedo

Federal University of Santa Catarina/PPGCIN, Florianópolis, Brazil
`gislaineparraf@gmail.com`, `priscila.bfagundes@gmail.com`,
`douglas.macedo@ufsc.br`

Abstract. The purpose of this paper is to present an analysis of the relationship between the principles of the Brazilian General Data Protection Act - LGPD and the stages of the data lifecycle. An analysis was made about of the objectives of each stage of the data lifecycle and the principles of the legislation. This analysis made it possible to answer the following research question: What is the relationship between each Brazilian LGPD principle and each of the stages of the data lifecycle? The data lifecycle presents the steps in which data act in a given process and relate to each other forming a chain of dependency between them. Based on the results, it was possible to observe that the data lifecycle model can be used to support and systematize the law compliance activities, since the law principles presented relations with the model stages. It was also observed that the principles of adequacy of data processing for the purposes of use and transparency are those that guide the other principles and that, besides these, the principles of safety, prevention, responsibility and accountability recommended by law, were related with all phases of the data lifecycle.

Keywords: Data lifecycle · General Data Protection Act · Data protection

1 Introduction

Technological advancement has enabled greater access to data and thus new ethical issues to be addressed within the legislation on this subject. This scenario prompted the approval, on August 14, 2018, of Brazilian Law No. 13.7091 - General Personal Data Protection Act[1] (LGPD), based on the General Data Protection Regulation[2] (GDPR), a standard that regulates the processing of personal data in the countries of the European Union. Brazilian LGPD is similar to GDPR and has been sanctioned to provide Brazil with competitive conditions in conducting international business, and because it is an

---

[1] http://www.planalto.gov.br/ccivil_03/_Ato2015-2018/2018/Lei/L13709.htm.
[2] https://eur-lex.europa.eu/eli/reg/2016/679/oj.

R. Mugnaini (Ed.): DIONE 2020, LNICST 319, pp. 79–88, 2020.
https://doi.org/10.1007/978-3-030-50072-6_7

extraterritorial law, an adequate adequacy of organizations is necessary, regardless of its line of business.

Brazilian LGPD provides for the processing of personal data, including in digital media, by a natural person or legal entity governed by public or private law, in order to protect their fundamental rights of liberty and privacy. According to the law, activities for the processing of personal data must comply with the following principles: purpose, appropriateness, necessity, free access, data quality, transparency, security, prevention, non-discrimination or responsibility and accountability [1].

These principles must be considered by organizations to adapt their processes, systems, and services and is necessary to define an approach to support this movement.

In order to provide support to organizations and assist them in complying with the legislation, an analysis was made of the principles proposed by the law and identified in which phases proposed by [2], for the data lifecycle, need to be observed. [2] proposes in his model that the data lifecycle be divided into four phases: collection, storage, retrieval and disposal.

If, on the one hand, the data lifecycle segments the process in phases according to its acting, on the other, the principles guide the data protection treatment activities. Thus, it is understood that the analysis presented in this study, by indicating the relationships that exist in each phase, can help organizations in the adequacy activity.

This article aims to present the results of this analysis and foster new studies involving Brazilian LGPD and the data lifecycle, including research involving the proposition of models to be used in the process of adequacy of organizations that deal in some way with storage and use of personal data to the new legislation.

2 Theoretical Foundation

2.1 Brazilian General Personal Data Protection Act (LGPD)

The Constitution of the Federative Republic of Brazil of 1988 recognizes the fundamental right to privacy and freedom of expression, which consequently guarantees the privacy of people's data and their right to instrumentalize them as their own [3]. However, globalization and the development of new technologies have fostered competition between organizations regarding the security of corporate information, as well as their customers. According to [4], companies and the government are increasingly vulnerable to espionage or malicious attacks that lead to information leakage or misuse. The Personal Data Protection Act comes to broaden the scope of protection of corporate information, including in this scenario the need to protect personal data.

Following a number of data misuse scandals in Europe and the United States involving large companies such as Cambridge Analytica and Facebook, governments in many countries have been forced to regulate access, storage, use and the dissemination of personal data of its citizens.

According to the current European GDPR on data protection, in its Article 4, personal data is characterized as a set of information concerning a living person that can lead to his/her identification and which should be protected regardless of how it is stored or processed, whether technologically or manually, provided they are organized according to predefined criteria [5]. According to the author, GDPR was based on a horizontal scope perspective, applying to companies from different areas that have contact with personal data by any means, just as it has extraterritorial coverage, inserting itself even outside the European Union.

According to [6], Brazilian LGPD was based on GDPR which gives people the right to have control over their data, and regulates business models based on the use of personal data. This regulation is the result of an adaptation of the European Data Protection Directive[3] adopted in 1995, which considered commercial use of the Internet from systems using personal data.

The GDPR has extraterritorial applicability and establishes that the circulation of personal data originating and destined in non-European Union countries must follow the same standards and security principles defined in European legislation, which has made other countries, among them Brazil, adapt to these issues and also establish its internal guidelines. Thus, Brazilian LGPD was approved to provide the country with greater security and privacy in personal data, enabling its insertion and its competitiveness in the world scenario.

The Brazilian LGPD in its Art. 6, determines a set of 10 principles that should guide the activities of processing of personal data. According to [6], by including such principles, Brazilian LGPD guarantees data subjects the right to request from public and private bodies information about how their data are used, which has a deadline to comply with the data owner's request. Processing activities for personal data should comply with the following principles [1]:

I - purpose: carrying out the treatment for legitimate, specific, explicit and informed purposes to the holder, without the possibility of further treatment incompatible with those purposes;

II - adequacy: compatibility of the treatment with the purposes informed to the proprietor, according to the context of the treatment;

III - need: limitation of the treatment to the minimum necessary for the accomplishment of its purposes, with the comprehensiveness of the relevant data, proportional and not excessive in relation to the purposes of the data processing;

IV - free access: guarantee to holders, free and easy consultation about the form and duration of treatment, as well as the completeness of their personal data;

V - data quality: guarantee to the owners, accuracy, clarity, relevance and updating of the data, according to the need and for the fulfillment of the purpose of its processing;

VI - transparency: guarantee to the holders of clear, accurate and easily accessible information about the treatment and its respective agents, observing the commercial and industrial secrets;

VII - security: use of technical and administrative measures capable of protecting personal data from unauthorized access and accidental or unlawful destruction, loss, alteration, communication or dissemination;

VIII - prevention: adoption of measures to prevent the occurrence of damage due to the processing of personal data;

IX - non-discrimination: the impossibility of carrying out the treatment for illicit or abusive discriminatory purposes;

---

[3] https://eur-lex.europa.eu/eli/dir/1995/46/oj.

X - Accountability: Demonstration by the agent of the adoption of effective measures capable of proving compliance with personal data protection rules, including the effectiveness of such measures.

Brazilian law follows GDPR guidelines on responsibilities for international transfer of personal data in defining the role of personal data processing agents (controller, operator and responsible for processing personal data). Likewise, rules are set regarding governance, good practices, security and data confidentiality. The Brazilian LGPD should be applicable even in the case of companies based abroad, provided that the data processing operation is performed in the national territory [7].

It is noteworthy that the law presents other directives in their content, but for the analysis that is proposed in this article, the 10 principles presented here were listed because it is understood that they guide the other items of the law.

2.2 Data Lifecycle

The data lifecycle comprises the steps in which data acts in a given process. Such steps correspond to the phases in which the data relate forming a chain of dependency between them. For [8] there are several types of lifecycles, as well as different stages associated with them, which can be evaluated from different perspectives and at different managerial levels. According to the author, for example, at the managerial activity level, the lifecycle is managed in the business process domain. At the project or system level, however, the data lifecycle is managed in the context of product or service development.

Regarding data, new research involving the theme has enabled various ways of collecting, storing and retrieving them. [2] proposes that the data lifecycle be divided into four phases, namely: collection, storage, retrieval and disposal and presents that these phases are permeated by six factors: privacy, integration, quality, copyright, dissemination and preservation as illustrated in Fig. 1.

The purpose of the collection phase is to meet the information needs and it is in it that the activities related to the initial definition of the data to be used are developed, as well as the planning of how they will be obtained, filtered and organized, identifying the structure, format and the means of description that will be used.

During the storage phase, activities related to processing, transformation, insertion, modification, migration, transmission and any action aimed at data persistence on a digital medium are performed. The focus of this step is to enable data reuse.

The recovery phase is related to the consultation and visualization of the data. Its purpose is to enable better access and use of them. In the disposal phase occurs the identification of data that is no longer needed and that can be deleted, focusing on eliminating unnecessary data.

Regarding the six factors that permeate the phases of the data lifecycle presented by [2], they are: privacy that deals with aspects that guarantee the privacy of people or institutions related to the data to be used; integration refers to the identification and use of requirements that will provide the integration of data with other data; the data quality, which is related to aspects such as origin, collection mechanisms, physical and logical integrity, among others to be considered to ensure that the data is reliable and useful; copyright is related to respect for the copyright of the data; dissemination that is linked

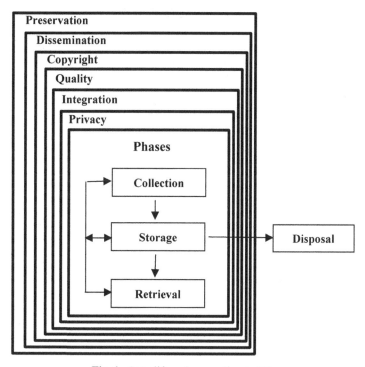

Fig. 1. Data lifecycle according to [2].

to findability and access to data; and data preservation that is related to preserving useful data so that it can be used in the future.

Table 1 presents a summary of the relationships defined and presented in [9], involving the four phases of the data lifecycle and the six factors that permeate them.

For [9], the lifecycle proposed by him aims to provide a structure that supports the efforts, studies and actions performed to obtain, maintain and use data, making it possible to approximate similar elements and distribute theories and methodologies depending on its scope, either by phases or by factors.

Table 1. Phases × factors involved in the data lifecycle process.

	Collection	Storage	Retrieval	Disposal
Privacy	Collect while respecting privacy	Store data by adopting access control	Recover data taking users into account the content to be made available	Discard the data respecting the right of privacy and request of the "owner" of the data
Integration	Collect data that can be integrated with other databases	Store data considering how to access and adopt a SGBD[a] that enables integration with other data	Recover data with the benefits of good integration that provides greater value in its use	Discard the data by observing the consequences of deletion in relation to content derived from integrations made with other data
Quality	Collect data considering the origin and the collection mechanisms used	Store data for its physical and logical integrity	Recover data with the same quality aspects as in the collection and storage steps	Discard the data by logging information about the deletion process so as not to impair the quality of future queries and usage
Copyright	Collect respecting the copyright of the author	Store the data by linking the source of data collection	Retrieve data by explaining usage permissions and how data can be used	Discard the data while maintaining information of its own to ensure legal compliance in derivative and/or referenced works of the discarded data
Dissemination	Collect data to support data findability and access	Store data by providing interpretable and easily accessible means of accessibility	Retrieve data considering elements and strategies that allow its location and access through collection processes	Discard data by looking at the impact of eliminating key search elements and finding datasets
Preservation	Collect data so that it can be preserved	Store with the premise that data can be interpreted in the future	Recover data with the possibility of obtaining the same interpretation at different times	Discard data from systems considering keeping a copy of deleted data to preserve it if there are unforeseen demands that require it

[a]SGBD - Database Management System.

3 Methodology

This study is considered a basic research as to nature, as it aims to generate new knowledge, useful for the advancement of science, with no practical application foreseen [10] and bibliographic as to procedures, which in the definition of [11], "is made from the survey of theoretical references already analyzed, and published by written and electronic means, such as books, scientific articles, web site pages".

As for the objectives, it is exploratory, as it aims to gain a better understanding of the problem to be studied and promote greater familiarity with the themes, to make them more explicit or build hypotheses [12]. It has a qualitative approach, since it is concerned with deepening the understanding of a social group, an organization, among others without considering numerical representativeness [10].

For the analysis of the relationships between Brazilian LGPD principles and data lifecycle stages, we used the data lifecycle model proposed in [2] and Brazilian LGPD principles for data processing. personal data present in the legislation. To identify such relationships, the following question was answered: What is the relationship between each of the Brazilian LGPD principles and each of the stages of the data lifecycle proposed by [2]?

The relationships were discovered considering that Brazilian LGPD principles can have the following relationships:

- Very Relevant: for situations where the principle directly influences the data lifecycle stage - represented by the symbol ●;
- Relevant: for situations where the principle indirectly influences the data lifecycle stage - represented by the symbol ○;
- Non-Identified/Unidentified: for situations where no relationships were detected - represented by the acronym NI.

The results of the analysis will be presented at the next session.

4   Analysis of the Brazilian LGPD Principles in Relation to the Stages of the Data Lifecycle

Table 2 shows the relationships identified and then the justifications for each of them are presented. The table outlines the Brazilian LGPD principles and the data lifecycle steps in the columns.

Table 2. Relationship between Brazilian LGPD principles and data lifecycle steps.

	Collection	Storage	Retrieval	Disposal
Goal	●	○	●	NI
Adequacy	●	●	●	●
Need	●	○	●	○
Free access	NI	●	●	NI
Data Quality	●	●	●	NI
Transparency	●	●	●	●
Safety	●	●	●	●
Prevention	●	●	●	●
Non-discrimination	NI	NI	NI	NI
Accountability	●	●	●	●

● Very Relevant: Directly Influences the Data Lifecycle Step.
○ Relevant: Indirectly Influences the Data Lifecycle Step.
NI: Non-Identified Relation.

The analysis identified that the **purpose** principle has a very relevant relationship with the *collection* and *recovery* stages. Considering that the *collection* has the goal of supplying the informational need, identifying the **purpose** of using the data is paramount to meet this objective. It was also identified that the **purpose** principle has a relevant relationship with the *storage* step, considering that it is necessary to store the necessary data for the identified **purpose**.

For the **adequacy** principle, a very relevant relationship was identified with all phases of the data lifecycle. Considering that this principle concerns the treatment of data appropriate to its purpose, it was observed that all phases have a direct and important relationship with this principle.

For the principle of **necessity**, a very relevant relationship was observed with the *collection* and *recovery* phases of the data lifecycle. Because it is a principle that recommends limiting treatment to the minimum necessary to achieve its purposes, it is observed that minimization in *collection* and *recovery* can be adopted. A relevant relationship with the *storage* and *disposal* phases has also been identified for this principle, considering that minimal data retention can be adopted and disposal should take into account the retention time required and consistent with the intended use of the data.

Regarding the principle of **free access**, a very relevant relationship with the *storage* and *retrieval* phases was identified, as these phases include access control in data storage, but allowing access and retrieval of the contents to be made available. For this principle, no relationships were identified with the collection and disposal phases.

Regarding the **data quality** principle, very relevant relationships were identified with the *collection, storage* and *retrieval* phases. The similarity of the quality factor acting in all phases of the data lifecycle with the principle of data quality favors the direct relationship identified: in the collection phase by defining elements that enable the perception of the quality of the collected data; storage phase definitions to ensure data maintains its physical and logical integrity; and the resources made available on recovery that should reflect these same aspects as described for collection and storage.

For the principle of **transparency**, a very relevant relationship was identified with all phases of the data lifecycle. Because this principle aims to provide holders with the guarantee of clear, accurate and easily accessible information on the processing of their data, this principle is directly associated with all phases, influencing the resources adopted in each of them.

By assessing the **security** principle, it was also possible to identify a very relevant relationship with the data lifecycle. It is noted that in order to protect personal data from unauthorized access and from accidental or unlawful situations concerning destruction, loss, alteration, communication or dissemination, technical and administrative measures must act from data collection to its disposal.

As with the principles of transparency and security, as regards the principle of **prevention**, a very relevant relationship with all stages of the data lifecycle has also been identified. It is considered that to prevent damage to personal data due to processing, it is necessary to adopt preventive measures at all stages since data damage can occur in any of them.

For the principle of **non-discrimination** no relationship was identified with any phase of the data lifecycle. It is understood that this principle is related to the definitions of the data use process and not to its lifecycle.

Regarding the principle of **accountability**, there was a relationship with all phases of the data lifecycle because to demonstrate the adoption of effective measures capable of proving compliance and compliance with personal data protection standards, these must be employed, practiced and validated at all stages of the cycle.

To exemplify the application approach idealized in this work, was considered the systems development process and the data collection phase. In this phase, the relevant relations with the principles of Purpose, Adequacy, and Necessity were identified. One of the alternatives to be adopted is about to the principle of "collecting the minimum necessary" guided by questions, such as: What is the data really needed for system scope? What is the purpose of data collection? What is the proper way to collect such data, considering its purpose?

Thus, it is observed that the relationships presented in this study and the reflections they provoke can support organizations in the task of adaptation to the Brazilian LGPD guided by the performance of the data.

The relationships identified in this analysis are limited to indicating in the phases of the data lifecycle which principles of the law to be addressed, and their purpose is not to present what should be done or how to implement the principles of the Brazilian LGPD. The analyzes presented can be used by organizations of any nature that aim to comply with the law or assess their compliance with the law, identifying at each stage of the data lifecycle which principles need to be implemented or need to be improved.

5 Conclusions

LGPD has been sanctioned in Brazil since August 2018 and regulates the protection of personal data, ensuring the exercise of personality rights, and sets limits on the right of access to third party information and the use of such data for discriminatory, unlawful or illegal purposes. The purpose of this study was to analyze the new law, considering that its guidelines ensure the effective protection of personal data, that organizations must adapt to it within a period of 1 year after its publication and to verify if this adequacy can be guided by data lifecycle steps.

As final considerations it can be concluded that the data lifecycle model proposed by [2] and the relations existing with the principles of the Brazilian LGPD can be used as a support tool for the process of compliance by organizations.

It can also be observed that the principles of adequacy, transparency, security, prevention, and accountability are the principles that have been related to all phases of the data lifecycle, but when evaluating the purpose of each principle, the Transparency and adequacy guide the others. Considering that transparency is one of the main goals of Brazilian LGPD, this and the adequacy of data processing for use purposes, direct the other principles.

As a suggestion of future work, it is proposed that a study be conducted to identify the actions that can be taken to adapt Brazilian LGPD principles at each phase of the data lifecycle considering the relations identified with the principles of the legislation.

References

1. Brasil. Lei n° 13.709, de 14 de agosto de 2018. Lei Geral de Proteção de Dados Pessoais (LGPD). Brasília (2018). Accessed 15 Sept 2019
2. Sant'ana, R.: Ciclo de vida dos dados e o papel da ciência da informação. In: Encontro Nacional de Pesquisa em Ciência da Informação, 14. UFSC, Florianópolis (2013)
3. Carvalho, L., et al.: Desafios de Transparência pela Lei Geral de Proteção de Dados Pessoais. In: Anais do VII Workshop de Transparência em Sistemas, pp. 21–30. SBC, Belem (2019)
4. Rapôso, C., et al.: LGPD - Lei Geral de Proteção de Dados Pessoais em Tecnologia da Informação: Revisão Sistemática. RACE Revista da Administração **4**, 58–67 (2019)
5. Costa, M.: A Era da Vigilância no Ciberespaço e os Impactos da Nova Lei Geral de Proteção de Dados Pessoais no Brasil: Reflexos no direito à privacidade. 2018. 93 f. TCC (Graduação) - Curso de Direito, Faculdade Nacional de Direito, Universidade Federal do Rio de Janeiro, Rio de Janeiro (2018)
6. Monteiro, R.: Existe um direito à explicação na Lei Geral de Proteção de Dados do Brasil? Instituto Igarapé, Artigo Estratégico 39. Rio de Janeiro, Brasil (2018)
7. Ferreira, D., Marques, R., Natale, A.: A política de informação na arena da privacidade dos dados pessoais. In: Encontro Nacional de Pesquisa em Ciência da Informação, 19. UEL, Londrina (2018)
8. Hernon, P.: Information lifecycle: its place in the management of US government information resources. Gov. Inf. Q. **11**(2), 143–170 (1994)
9. Sant'ana, R.: Ciclo de vida dos dados: Uma perspectiva a partir da Ciência da Informação. Revista Informação Informação **21**(2), 116–142 (2016)
10. Gerhardt, T., Silveira, D.: Métodos de pesquisa. UFRGS, Porto Alegre (2009)
11. Fonseca, J.: Metodologia da pesquisa científica. UEC, Fortaleza (2002)
12. Gil, A.: Como elaborar projetos de pesquisa, 3rd edn. Atlas, São Paulo (1991)

A Method for Collecting Provenance Data: A Case Study in a Brazilian Hemotherapy Center

Márcio José Sembay(✉) ⓘ, Douglas Dyllon Jeronimo de Macedo ⓘ, and Moisés Lima Dutra ⓘ

Federal University of Santa Catarina, Florianópolis, Brazil
marcio.sembay@posgrad.ufsc.br,
{douglas.macedo,moises.dutra}@ufsc.br

Abstract. Data provenance is a process that aims to provide an overview of the origin of data used by information systems. It focuses on the origin of the data, especially on identifying the data sources and the transformations the data has undergone over time. This paper proposes a method for data collection based on the Provenance Model (PROV-DM), to be applied on Brazilian hemotherapy centers. Storing data on anemia indices using data provenance is the overall purpose of it. This work uses concepts of data provenance, knowledge provenance and scientific workflow techniques. It is an exploratory research, of practical and deductive nature, with application of a case study. Actual data was extracted from reports generated by a Brazilian hemotherapy center, provided from 2000 to 2018. People unsuitable for blood donation, who had favorable anemia rates to be rejected, were quantified and analyzed. A total of 197,551 blood donor candidates who attended the hemotherapy center in 19 years were analyzed. In the end, it was possible to quantify the unfit candidates with the highest index of anemia. A total of 1,011 male and 4,039 female candidates were accounted for, totaling 4.02% and 16.09% respectively of donors unfit for blood donations.

Keywords: Data Provenance · Anemia · Hemotherapy Center

1 Introduction

Information Science has its own scientific status as a social science. By its interdisciplinary nature, it presents interfaces with Mathematics, Logic, Linguistics, Psychology, Computer Science, Production Engineering, Graphic Arts, Communication, Librarianship, Administration, and other similar scientific fields [3]. Regarding the use of Data Provenance, both the Information Science and Computer Science use the structures of scientific workflows, which are abstractions related to the source of data, used as a support in the modeling of scientific experiments. Provenance is related to the audit, screening, lineage, and source of data. It can also be considered a metadata that describes the origin and all path taken to achieve the results of an experiment [10, 17].

R. Mugnaini (Ed.): DIONE 2020, LNICST 319, pp. 89–102, 2020.
https://doi.org/10.1007/978-3-030-50072-6_8

This paper proposes a method for collecting Provenance data related to anemia indices. According to specialists of a particular hemotherapy center in Brazil (which due to privacy conditions, will be reported here as "X Hemotherapy Center"), anemia is a generic name for a series of conditions characterized by deficiency in hemoglobin concentration or in the production of red blood cells. Hemoglobin is a blood element with the function of carrying oxygen in the lungs to nourish all cells in the body.

A current study shows that 30% of the world's population is anemic, especially children under 2 years old and women of different age groups, although it can also occur in men and the elderly. In addition, it is estimated that 27% to 50% of the population is affected by iron deficiency, especially in lower income and developing populations. In Brazil, the data may vary according to the study and the population group analyzed. But overall, it is estimated that 40% to 50% of children have anemia [25]. In this sense, it is important to emphasize that the data contained in the original database of the health institution under study in this paper do not establish systematic relationships between the stored variables for possible analysis of the statements generated by the specialists in the process of refusal of blood donors, regarding anemia rates.

There is no computational analysis of stored variables to uncover anemia index donations to chart possible future preventions, only expert-generated statements. These statements recorded by biomedical specialists do not always agree with database variables for possible broader analysis. Incorporating expertly defined statements regarding anemia rates, the possibility for obtaining a higher quality reduced dataset is evident. The analysis performed on the reduced dataset provided more reliable answers about a given biological phenomenon.

However, it was important to create a framework, which was able to facilitate the proposition of the method for data provenance activities and the storage of the expert statements, through an auxiliary database. Thus, the proposed method can ensure that statements made by experts during the process of blood donor refusal for anemia are reliable for new information flows and for the generation of new knowledge.

The proposed method is based on an adaptation of the Provenance Data Model (PROV-DM), which consists of a computational strategy capable of ensuring that expert-generated statements are passed on from the original X Hemotherapy Center database to an auxiliary database. Its main goal is to provide for a broader analysis and improve the quality of blood donations. In the end, the proposed method was able to manage the statements generated by specialists during the process of blood donor refusal for anemia indices, which were obtained from reports generated by the X Hemotherapy Center database from 2000 to 2018. The structuring of the method is based on data provenance stages, the use of scientific workflows and the needs found throughout the research for the treatment of digital data.

This research was developed in the scope of the Doctoral Program in Information Science from the Federal University of Santa Catarina, Brazil.

2 Literature Review

2.1 Information Science

Information Science is an area that is directly or indirectly linked to information technologies in the use of its methods of organization and representation of information in research development. Information technologies are key elements in the development of Information Science, as the creation of technological tools promotes the development of theories in order to achieve the goals set by this science in relation to the problems it is dedicated to solving [19].

The focus of Information Science implies both sociological and epistemological approaches, focused on phases such as: generation, collection, organization, interpretation, storage, retrieval, dissemination, transformation, and use of information. Information Science is an interdisciplinary science that assumes several disciplines of technological knowledge and try to contribute to the generation of new scientific knowledge [3, 7].

2.2 Anemia in Blood Donors

In 1999, members of the United Nations International Children's Emergency Fund (UNICEF), the United Nations University (UNU), the World Health Organization (WHO) and the Micronutrients Initiative (MI) showed that 3.5 billion people worldwide have iron deficiency anemia and that iron deficiency may be present in 80% of the world's population [23].

One of the most frequently observed factors in assessing the presence of anemia in blood donor candidates in hemotherapy centers throughout Brazil is hematocrit levels, the percentage of volume occupied by red blood cells and hemoglobin, which are the main items extracted from the statements in the original X Hemotherapy Center database, along with the fit and unsuitable candidates for blood donations and their respective screening. In Brazil, Ordinance RDC 153 of July 2004, enacted by the Ministry of Health, establishes the minimum acceptable hemoglobin and hematocrit values for a blood donation. These values are: 13 g/dl hemoglobin and 39% hematocrit for men and 12.5 g/dl hemoglobin and 38% hematocrit for women [13].

In the same Ordinance, the Ministry of Health also determined the minimum interval that must be respected between blood donations. This interval should be eight weeks for men and twelve weeks for women, respectively, because the shorter the interval, the greater the chance of developing anemia [13].

2.3 Scientific Workflows

Scientific experiments consist of observing a phenomenon through data analysis, and using the results obtained to prove or disprove a hypothesis. Due to the need to organize, process, control, and analyze the experiment, its representation is made through a cycle whose steps are composition, execution, and analysis. A scientific workflow is an abstraction of this process, which allows the formal specification of the steps to be performed in a scientific experiment [11].

An example of using scientific workflows would be to capture the steps taken to create a new drug, i.e. the source of the data that led to the creation of such a formula. In this sense, whenever this formula was improved, we would have the original data for reuse and replication of the experiment.

To benefit from provenance data, this data has to be captured, modeled, and stored for future reference. Information on the provenance of stored data can be managed by various Scientific Workflow Management Systems (SGWfCs) [15, 16]. Some SGWfC, such as Taverna [18], Kepler [2] and Pegasus [11] allow you to capture workflow steps during their execution. However, these systems often adopt proprietary models to capture the provenance generated in executions [8].

In this paper, a specific workflow was developed to demonstrate data capture using an auxiliary database without the need to change the original database.

2.4 Data Provenance

Data provenance is the complementary documentation of a given data that contains the description of "how", "when", "where", and "why" it was obtained and "who" obtained it [5]. When buying a work of art, it is important to know its origin from its inception, including all former owners, i.e. this information will be essential to establish the value of this work of art. The same is true of data where data provision makes it possible to ensure data quality and accuracy [21].

In this sense, whenever provenance is automatically captured, it can be divided into levels [10]: a) workflow: involves the execution description of a process, i.e. the tasks that are part of it, is used by the vast majority of solutions with SGWfC and, in this case, must be adapted to capture the data from the different processes executed; b) activity: can occur in two ways. In the first, each executed process/program changes to capture the provenance data. In the second, specific programs can be created to monitor the execution of a given process and capture the provenance data; c) operating system: uses the data provided by the system, storing it in a specific database for provenance analysis.

By using data provenance it is possible to keep a complete record of how the calculation or processing was performed and it is essential to [6]: (a) ensure repeatability, (b) catalog the result, (c) avoid duplication of effort, and (d) retrieve data sources from output data. The main benefits of provenance for data quality are [4]: a) communicates data quality: reliability, suitability, accuracy, timeliness, redundancy; b) improves data interpretation as a function of source recognition; c) contributes to the justification of the use of a given data; d) reduces the possibility of errors in judging the accuracy of the data; e) allows non-data expert users to understand the processing steps; (f) identify the process used to conduct the creation of scientific data; g) allows updating of data from relational views; h) allows modification of relational view schemas; i) allows the use of historical data sources.

In this sense, the application of data provenance can be observed in the most varied areas, such as digital libraries, food industry, journalism, the traceability of information in social networks and the transparency of commercial applications, among others [9].

Provenance of Knowledge

The term provenance of knowledge includes the source of the so-called meta-information, which is based on obtaining a description of the origin of part of the knowledge, including a description of the reasoning method used to generate it. However, data provenance and knowledge provenance have the same concerns and motivations, differing as to the purpose of the record that will be captured [20].

The provenance of knowledge provides two aspects: a) a personal and more abstract view of a document and its derivations, specifically for the experiment and the person, with the direct contribution of the scientist; and b) a more specific understanding of the data processing domain or its execution process, and may receive contributions from both the scientist and the note-taking curators [22, 23].

In this work, provenance of knowledge is related to the context of observing the statements on anemia rates described by the experts in the reports provided by the X Hemotherapy Center on donors who have become unfit for blood donation, determining the reliability of the researcher's reasoning about a given dataset. The provenance of knowledge was a term used to demonstrate and record the rules and reasoning used in the sample derivation processes from the reduced dataset, obtained at the X Hemotherapy Center X, in relation to the data relating anemia rates in unfit blood donors.

3 Related Works

In a PhD thesis written in 2012 at the University of São Paulo, the author proposes a model for describing data provenance for knowledge extraction in hemotherapy information systems based on the Open Provenance Model (OPM), designed to manage provenance records. Other similar applications can be found in the paper entitled "Laboratory and clinical genomic data sharing is crucial to improving genetic health care: the position statement of the American College of Medical Genetics and Genomics" [12]. In this research, the institution responsible for the study presents clinical level patterns by which statements about gene/disease associations and the clinical significance of variants were captured, by means of data provenance techniques, in statements made by experts in shared genomic data systems.

4 Proposal

This paper proposes a method for collecting provenance data related to anemia indices, by adapting some components taken from PROV-DM. PROV-DM's main function is to describe people, entities, and activities involved in the production of data. In addition, the PROV-DM model provides the conditions for provenance to be demonstrated and exchanged between different systems. For this purpose, a data provenance application was created. This application uses an auxiliary database to store provenance data related to anemia indices. It abstracts some attributes from the original database through researcher analysis. When searching the database, the provenance process assists in the process of tracking and signaling from the data source, as well as their movement between different data sources [21].

The proposed method sought to store the statements related to amounts of blood donation candidates with anemia rates considered unfit for blood donations, taken from reports provided by the X Hemotherapy Center system, in Brazil, from 2000 to 2018. The X Hemotherapy Center board has provided 19 years of reports from its blood donation registration system, containing various attributes and at least 80 reasons for refusing blood donors. All information provided has preserved the confidentiality of blood donors.

These reports were transformed into a CSV file by a software application created with the Java Eclipse IDE. Also, a local PostgreSQL database was created to serve the auxiliary database, i.e. the local repository of provenance. This local repository was populated with records taken from the original database provided by the X Hemotherapy Center X. Without changing the structure of the original database, It was possible, to observe the predominance of anemia indices found in blood donors at the X Hemotherapy Center X during 19 years.

Finally, the workflows were generated with Taverna Workbench Core, and the Jasper-Reports library was used to generate a report that was, subsequently, transformed into a reduced dataset (see Table 1).

4.1 Case Study

PROV-DM is divided into six components that contain both the elements and the possible relationships between them. They are [24]: a) Entities and Activities. Entities can represent any object (real or imaginary) and Activities represent the processes that use and generate Entities; b) Agent and Responsibilities: Agents are Entities that influence, directly or indirectly, the execution of the Activities, receive attributions from other Agents and may have some kind of connection (ownership, rights, etc.) over other Entities; c) Derivations: Describes the relationship between different Entities during the transformation cycle performed by the Activities, allowing to demonstrate the dependency between the used and generated Entities; d) Alternative: Describes the relationship between different views of the same Entity; e) Collections: These are Entities that have members, which are also Entities, and may have their provenance shown collectively; f) Annotations: Provides mechanisms for adding annotations to elements of the model.

Figure 1 presents the proposed method. Four steps are proposed in order to accomplish the task: i) Entities; ii) Activities, Agents, and Derivations; iii) Alternative Collections; and iv) Notes.

These steps comprise the adaptation of some PROV-DM components for provenance management, the creation of a specific workflow for data extraction, and the analysis of 19 years of anemia indices found in candidates who were considered unfit for blood donations.

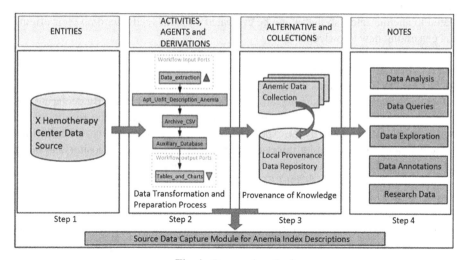

Fig. 1. Proposed method

To better explain the proposed method presented in Fig. 1, the following subsections present each step with its respective components.

Step 1 (Entities)
Data provided by the X Hemotherapy Center data source are collected and selected so that they can undergo the second step of this transformation process.

Step 2 (Activities, Agents, and Derivations)
In this step, "ACTIVITIES, AGENTS, and DERIVATIONS" are represented by a workflow created specifically for the cycle of activities required to prepare the data collected before to be manipulated and studied, regarding the statements reported by the experts. Here the workflow aims to extract data that will be seen as reliable to perform analysis for blood donor improper anemia indices (rejections) and, soon after, the extracted data is transformed into a CSV file for the creation of an auxiliary database to generate tables and graphs (data preparation and transformation process).

So, the "ACTIVITIES" are represented by the dataset collected from the X Hemotherapy Center over 19 years, containing dates, times, consumption, processes, transformation, modification, relocation, and use of the original data in relation to anemia rates. The activities are performed by the agents. The "AGENTS" are represented by both blood donor candidates and health specialists who appear in X Hemotherapy Center reports. The "DERIVATIONS" represents the transformation of the X Hemotherapy Center, after the application of the method proposed here, in an entity that will serve as a reference model for the application of this method in other Brazilian hemotherapy centers with the same structure.

The second step starts with the workflow created specifically for the process performed at the X Hemotherapy Center, which is the process of collecting provenance data related to anemia rates, so that it can be evaluated and analyzed regarding their origin, thus generating new knowledge.

Step 3 (Alternative and Collections)
"ALTERNATIVE" represents the view of the X Hemotherapy Center in the declarations of anemia indices of candidates for blood donation considered inept. By declaring the reports provided, it was possible to quantify in order to generate the results of the analysis of the data declared by the experts. "COLLECTIONS" represents the collection of anemic candidate data to be inserted into the local source data repository. In here, one can also apply provenance of knowledge concepts for possible audits of source data. It is important to notice the "COLLECTIONS" component stores data that constitute documents, in which each document has its own provenance, but the file itself also has its origin information: who kept it, which documents contained it at what time, how it was assembled, etc. Therefore, in addition to the procedures for collecting the provenance data, i.e. the provenance of knowledge applied to this method, it was possible to provide an overview of the provenance of the data described in the reports on the anemia indices of candidates for unfit donations. This made possible a better understanding of the risks and the reasons for the rejection of the blood donations.

Step 4 (Notes)
After storing the provenance data in a local repository, in the fourth step, "NOTES" represents the relationship of the important points on anemia indices, such as: analysis, consultations, exploration, annotations, and reuse of data for further research. In here, it is possible to generate reports that can be cross-referenced with other data, as needed by health specialists. Consequently, this step creates an interaction between refined data and expert reporting.

5   Results

The reports provided by the X Hemotherapy Center are annual from 2000 to 2018, dated January 1 to December 31 of each year to better simplify and reduce the presentation of the data during these 19 years. The body of each report contained a series of attributes, from which only the data that had the potential to generate the expected results was selected.

The selected attributes were: i) number of male and female fit donors as well as the number of male and female unfit donors, all aged from 16 to 60 years old or older, including first-time, repeat or sporadic donors; ii) anemia indices of male and female unfit donors, i.e. low hematocrit and low hemoglobin, considered the reasons for refusal according to statements made by experts in the submitted reports, which in fact built the set of information necessary to generate the process of data provenance and provenance of knowledge; and finally iii) the screening performed by the experts in each year surveyed, i.e. the discovery of diseases through blood donation at the X Hemotherapy Center. These attributes can be better observed in Fig. 2 below.

	ano [PK] integer	indanemfem integer	indanemmasc integer	quantaptofem integer	quantaptomasc integer	quantinapfem integer	quantinapmasc integer	trimed integer
1	2000	647	237	1608	4418	1612	2307	71
2	2001	876	278	1674	4316	1844	2409	0
3	2002	636	203	1974	4313	1852	2328	0
4	2003	269	65	2006	4332	1267	1767	0
5	2004	188	46	1928	3970	1447	1962	0
6	2005	99	16	2209	4285	1265	1484	0
7	2006	102	13	2359	4667	1505	1383	0
8	2007	123	13	2716	4722	1349	1152	0
9	2008	146	28	3021	4899	1263	1091	0
10	2009	167	25	2905	4919	1401	1142	0
11	2010	158	32	3116	4886	1336	1091	46
12	2011	59	8	3672	5362	1318	991	114
13	2012	109	3	3886	5176	1232	861	19
14	2013	102	9	4019	5165	1124	813	74
15	2014	2	0	4506	5516	1194	929	17
16	2015	10	1	4924	5783	1241	1047	12
17	2016	3	2	3943	4527	1011	842	1
18	2017	137	16	3475	4021	928	723	19
19	2018	206	16	3692	4380	916	834	381

Fig. 2. Auxiliary database with selected attributes.

The auxiliary database presented in Fig. 2 demonstrates the selected attributes taken from 19 years of reports provided by the X Hemotherapy Center. They are a massive set of information that was possible to be retrieve and group into a reduced dataset in order to be analyzed.

The attributes selected are the following (in Brazilian Portuguese acronyms): a) ano (donation reference year); b) indanemfem (female anemia index); c) indanemmasc (male anemia index); d) quantaptomasc (amount of male fit donors); e) quantaptofem (amount of female fit donors); f) quantinapfem (amount of female unfit donors); g) quantinapmasc (amount of male unfit donors); and finally h) trimed (screening by the attending physicians).

Table 1, shown below, is populated with data extracted from the outcome of the analysis undertaken on the auxiliary database. This database contains statements about the anemia indices of unfit donors. All the attributes shown in Table 1 helped perform the development of the specific workflow to create the provenance data method for the X Hemotherapy Center.

Table 1 shows the number of eligible and unsuitable blood donation candidates, both male and female, as well as the percentage of anemic unsuitable blood donation candidates. Importantly, candidates unfit for blood donations were rejected for anemia rates, i.e. were also filtered from at least 80 reasons for refusal before becoming eligible for blood donations.

These reasons for refusal are diverse, ranging from something simple as a fever, weight loss, flu manifestations, among others, to more complex reasons such as diabetes, heart disease, cancer, HIV, etc. In this work, we gathered the unfit donors who had several reasons for refusal. From this subset, we extracted those unfit by anemia, by presenting that percentage for each year. Table 1 also shows the amount of screenings performed by doctors each year, which in other words means the number of blood donation candidates who discovered disease during the donation process.

The 19 years of data analyzed revealed 197,551 blood donation candidates. Out of this total, 114,813 were male and after blood tests 89,657 were found fit and 25,156 were found unfit for blood donations. Anemia rates were present in 1011 candidates, totaling 4.02% of candidates unfit for blood donations. Of the remaining 82,738 female blood donation candidates, 57,633 became fit and 25,105 became unfit for blood donations after blood tests. Besides, there were 4,039 anemia inducing female candidates, totaling 16.09% of donors unfit for blood donations.

Table 1 also shows that from 2001 to 2009 there was no screening, but anemia rates continued to fluctuate between male and female donors, with a female predominance. In 2018, the highest number of tests was observed, evidence for some unidentified specific reason in relation to blood donations that year. The year 2001 is the year in which the anemia rates between men (11.54%) and women (47.51%) represent the highest rates observed. If compared to 2018, when there was the highest screening rate, the anemia rate was well below the 2001 average.

Table 1. Number of donors (fit, unfit, anemic unfit, percentage of anemic unfit and screening).

Years	Male fit	Male unfit	Anemia unfit male	% Anemics unfit male	Female fit	Female unfit	Anemia unfit female	% Anemics unfit female	Screenings
2000	4418	2307	237	10,27%	1608	1612	647	40,14%	71
2001	4316	2409	278	11,54%	1674	1844	876	47,51%	0
2002	4313	2328	203	8,72%	1974	1852	636	34,34%	0
2003	4332	1767	65	3,68%	2006	1267	269	21,23%	0
2004	3970	1962	46	2,34%	1928	1447	188	12,99%	0
2005	4285	1484	16	1,08%	2209	1265	99	7,83%	0
2006	4667	1383	13	0,94%	2359	1505	102	6,78%	0
2007	4722	1152	13	1,13%	2716	1349	123	9,12%	0
2008	4899	1091	28	2,57%	3021	1263	146	11,56%	0
2009	4919	1142	25	2,19%	2905	1401	167	11,92%	0
2010	4886	1091	32	2,93%	3116	1336	158	11,83%	46
2011	5362	991	8	0,81%	3672	1318	59	4,48%	114
2012	5176	861	3	0,35%	3886	1232	109	8,85%	19
2013	5165	813	9	1,11%	4019	1124	102	9,07%	74
2014	5516	929	0	0,00%	4506	1194	2	0,17%	17
2015	5783	1047	1	0,10%	4924	1241	10	0,81%	12
2016	4527	842	2	0,24%	3943	1011	3	0,30%	1
2017	4021	723	16	2,21%	3475	928	137	14,76%	19
2018	4380	834	16	1,92%	3692	916	206	22,49%	381
Total	**89.657**	**25.156**	**1.011**	**4,02%**	**57.633**	**25.105**	**4.039**	**16,09%**	**754**

These comparisons help draw estimates and thresholds for new studies in the area of hemotherapy from a data provenance perspective, which could ultimately contribute to the prevention of anemia rates in the X Hemotherapy Center. In order to provide a more detailed comparison, Fig. 3 shows a graph of the profile of anemic unsuitable donor candidates during the 19 years analyzed.

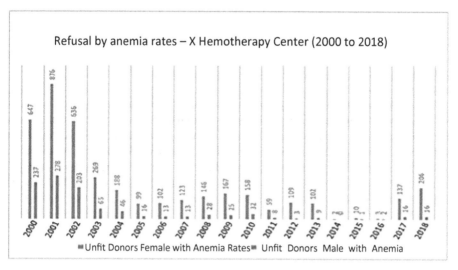

Fig. 3. Refusals due to anemia (male and female unfit)

It can be observed that men's anemia rates are lower than women's. This can be demonstrated as a result of the interval between blood donations from women and men. Actually, the X Hemotherapy Center has reported that women tend to be at risk of developing anemia earlier than men. This happens due to more frequent blood donations done by women in a shorter period of time, or even repeated donations.

Other factors that are associated with higher rates of anemia in women are pregnancy and the monthly blood loss from menstruation. For men, the possibilities are that continued blood loss caused by some type of bleeding, associated with blood donation, or regular blood donation may be related to the risk of developing anemia [1]. According to the literature, the recommendations to reduce anemia rates in all male and female population are related to the iron supplementation after blood donation and the increasing of the waiting time between donations [13, 14].

6 Conclusions

This paper proposed a method for data collection in a Brazilian hemotherapy center, based on the data provenance approach. This method proved to be important for generating a reduced dataset, in order to allow knowledge extraction and mining of a volume of data generated over a 19-year period.

The application of the concept of data provenance along with the provenance of knowledge and the scientific-workflow techniques for the development of the proposed method led to the conclusion that these elements together may, in fact, contribute to the advancement of research in Brazilian hemotherapy centers. They helped guide data collection in relation to the anemia index statements found in the 19 years of data presented in the reports provided by the X Hemotherapy Center. Moreover, they contributed to generate knowledge through the analysis performed in the anemia rates found. It could be observed that most of the population studied was female, who develops more frequently anemia rates in conjunction with other factors after blood donations, which may be a research factor for the discovery of other diseases. The method proposed here can be applied to another Brazilian hemotherapy center that has the same features and security policies presented here.

The main contributions of the proposed method are: a) the improvement of the analysis that discovers anemia indices that make blood donation unfeasible; b) the provision of a beneficial view of donor groups in their hemotherapy centers; c) the raising of the quality of blood donations by creating preventive mechanisms for blood donor evasion; d) the permission, when necessary, to query the local source data repository, in order to create data-quality metrics and to perform an audit processes for this data. These contributions demonstrate that the use of data provenance together with the provenance of knowledge are differential requirements of what can be found in the literature and, in fact, contribute to the relevance of the results found in this research.

After several searches in the literature and considering the few works found, it became clear to us the relevance of the studied subject, since we could not find similar proposals. We believe that methods for data collection that are able to highlight, synthesize and explain elements in the area of Hemotherapy may be a guide for the development of new research in this area.

It also became clear that Computer Science approaches, such as the Provenance of Data, combined with an Information Science viewpoint could be very useful to the context of hemotherapy information systems. Computer Science provides the technological support for the development of data provenance. On the other hand, Information Science provides the methods and techniques for informational treatment, making use of technological applications to apply the provenance of knowledge.

7   Future Perspectives

Some paths can be envisaged in the follow up of this research: a) Data collection done directly on the hemotherapy center's database with the help of an already-connected auxiliary database (a local provenance data repository composed of anemia declarations). This would be done without modifying the original database structure, i.e. by generating an automatic CSV file (or others) in a cloud computing structure capable of handling large amounts of data, providing better data quality for future research; b) Automating the data description process, i.e. generating predictions for more complex analysis by using data provenance as a preventive factor for anemia indices that result in blood-donation refusals; c) Improving the data provenance method by performing data-cross-referencing processes (reasons for refusal of blood donations in all hemotherapy centers in Brazil);

and d) Integrating more hemotherapy centers in this research, or even performing the study in other regions of Brazil, by adapting the proposed method whenever is necessary. Another future perspective means using data provenance for preventing anemia problems, by indicating the reasons why they occur more frequently in certain regions of Brazil. We can also envisage the evaluation of how data provenance models could assist in generating more complex and complete analysis, in order to discover the most prominent anemia rates, by Brazilian region.

For all of the aforementioned scenarios to become true, we believe that Brazil should improve its data storage infrastructure and computational tools applied to the Brazilian hemotherapy centers. Furthermore, it would be advisable to think about the creation of research institutes all over the country, in order to study and to prevent anemia rates and other reasons for refusing blood donations. That indeed would be a challenge.

References

1. Almeida, F.N.: Descrição da Proveniência de Dados para Extração de Conhecimento em Sistemas de Informação de Hemoterapia, p. 114 (2012). f. Tese (Doutorado) - Curso de Bioinformática, Bioinformática, Universidade de São Paulo - USP, São Paulo (2012)
2. Altintas, I., Berkley, C., Jaeger, E., Jones, M.: Kepler: an extensible system for design and execution of scientific workflows. In: Proceedings of 16th International Conference on Scientific and Statistical Database Management, Santorini Island, Greece, 23 June 2004, pp. 423–424. IEEE (2004)
3. Borko, H.: Information science: what is it? Am. Doc. **19**(1), 3–5 (1968)
4. Bose, R., Frew, J.: Lineage retrieval for scientific data processing: a survey. ACM Comput. Surv. **37**(1), 1–28 (2005)
5. Buneman, P., Khanna, S., Wang-Chiew, T.: Why and where: a characterization of data provenance. In: Van den Bussche, J., Vianu, V. (eds.) ICDT 2001. LNCS, vol. 1973, pp. 316–330. Springer, Heidelberg (2001). https://doi.org/10.1007/3-540-44503-X_20
6. Buneman, P., Tan, W.C.: Provenance in databases: tutorial outline. In: Proceedings of ACM SIGMOD International Conference on Management of Data, Beijing, China, 11–14 Jun 2007. ACM (2007)
7. Capurro, R., Hjorland, B.: O conceito de informação. Perspectivas em Ciência da Informação, Belo Horizonte **12**(1), 148–207 (2007)
8. Cuevas-Vicenttin, V., Dey, S., Wang, M.L.Y., Song, T., Ludäscher, B.: Modeling and querying scientific workflow provenance in the D-OPM. In: Proceedings of 2012 SC Companion High Performance Computing, Networking, Storage and Analysis, Washington, EUA, 10–16 November 2012, pp. 119–128. IEEE (2012)
9. Curbera, F., Doganata, Y., Martens, A., Mukhi, N.K., Slominski, A.: Business provenance – a technology to increase traceability of end-to-end operations. In: Meersman, R., Tari, Z. (eds.) OTM 2008. LNCS, vol. 5331, pp. 100–119. Springer, Heidelberg (2008). https://doi.org/10.1007/978-3-540-88871-0_10
10. Davidson, S.B., Freire, J.: Provenance and scientific workflows: challenges and opportunities. In: ACM SIGMOD International Conference on Management of Data, pp. 1345–1350 (2008)
11. Deelman, E., Gannon, D., Shields, M., Taylor, I.: Workflows and e-science: an overview of workflow system features and capabilities. Future Gen. Comput. Syst. **25**(5), 528–540 (2009)
12. Genetics in Medicine: ACMG. https://www.nature.com/articles/gim2016196. Accessed 22 Sept 2019

13. Mendrone, A.J.R., et al.: Anemia screening in potential female blood donors: comparison of two different quantitative methods. Transfusion **49**, 662–668 (2009)
14. Meyers, D.G.: The iron hypothesis: does iron play a role in atherosclerosis? Transfusion **40**(8), 1023–1029 (2000)
15. Moreau, L., et al.: The open provenance model core specification (v1.1). Future Gen. Comput. Syst. **27**(6), 743–756 (2011)
16. Moreau, L., Freire, J., Futrelle, J., McGrath, R.E., Myers, J., Paulson, P.: The open provenance model: an overview. In: Freire, J., Koop, D., Moreau, L. (eds.) IPAW 2008. LNCS, vol. 5272, pp. 323–326. Springer, Heidelberg (2008). https://doi.org/10.1007/978-3-540-89965-5_31
17. Moreau, L., Groth, P.: Provenance: An Introduction to PROV. Synthesis Lectures on the Semantic Web: Theory and Technology, vol. 3, no. 4, pp. 1–129. Morgan & Claypool Publishers, California (2013)
18. Oinn, T., Li, P., Kell, D., Goble, C.: Taverna/^{my}Grid: aligning a workflow system with the life sciences community. In: Taylor, I.J., Deelman, E., Gannon, D.B., Shields, M. (eds.) Workflows for e-Science, pp. 300–319. Springer, London (2007). https://doi.org/10.1007/978-1-84628-757-2_19
19. Saracevic, T.: Ciência da Informação: origem, evolução e relações. Perspectivas em Ciência da Informação **1**(1), 41–62 (1996)
20. Silva, P.P., Mcguinness, D.L., Mccool, R.: Knowledge provenance infrastructure. Proc. IEEE Data Eng. Bull. **25**, 179–227 (2003)
21. Simmhan, Y.L., Plale, B., Gannon, D.: A survey of data provenance techniques. Technical report TR-618, Computer Science Department, Indiana University (2005)
22. Stevens, R., Zhao, J., Goble, C.: Using provenance to manage knowledge of in silico experiments. Brief. Bioinform. **8**, 183–194 (2007)
23. Stolzfus, R.J.: Defining iron deficiency anemian public health terms: a time for reflection. J. Nutr. **131**, 565S–567S (2001)
24. W3C: PROV-DM. http://www.w3.org/TR/prov-dm/. Accessed 21 Sept 2019
25. WHO. https://www.who.int/topics/anaemia/en/. Accessed 21 Sept 2019

Personal Data Protection and Its Reflexes on the Data Broker Industry

Guilherme Birckan[✉], Moisés Lima Dutra, Douglas D. J. de Macedo, and Angel Freddy Godoy Viera

Universidade Federal de Santa Catarina, Florianópolis, Brazil
gbirckan@gmail.com

Abstract. Demonstrates the relationship between government and private interests in identifying people's profiles on the Internet. Describes the establishment and development of information aggregators and merchants, the data brokers. Discusses the boundaries of personal data commoditization, which in consequence wears away privacy and anonymity. Associates the inception of laws that mandate publicity to data breaches events, exposing the model, and ensuing debates on the need for further regulation. Presents the innovative generation of legislation created to govern a business that up until then operated free from public scrutiny. Introduces ideas to prevent the extinction of such a business model upon the shift to privacy and data protection.

Keywords: Digital identity · Internet privacy · Personal data · Data brokers

1 Introduction

Big Data, Cloud Computing, Cloudlets, Internet of Things, Data Brokers, all contemporary terms, buzzwords associated with what is being hyped in tech trends - except for the last one. Data brokers, although increasingly subject to debates, are as old as the net itself. After all, since information started being published online, there has always been a need for aggregators, as there always have been those who were interested in their products: compilations from sparse sources generating specific dossiers about something or someone.

There is power in identity, as according to Castells (2011), it "is people's source of meaning and experience". The most important link in the "who-what-where-when" tetrad, the unique identification of a being that is extracted from a data bulk (especially the Web) has been, for decades, the desire of states and corporations. Prins (2006) points out that "a look at our contemporary, data-based society reveals that information about people is essential for a variety of economically and socially useful and crucial purposes: education, taxation, social benefits, health care, crime detection and terrorism prevention, commerce and marketing, to name but a few".

On one side, ordinarily control-avid governments, armed with the national security argument; on the other, the refinement of targeted marketing, progressively individualized to declared and inferred tastes; in the middle, lies the citizen, the user, the target,

R. Mugnaini (Ed.): DIONE 2020, LNICST 319, pp. 103–117, 2020.
https://doi.org/10.1007/978-3-030-50072-6_9

fooled in that a mouse and a screen would reflect some degree of anonymity, while the truth is that every search, every click, every like, every post, are all cataloged, reunited, processed, and, not rarely (or maybe very frequently), sold to third parties.

Usual entry points for data acquirement, and at first sight somehow innocent, are the filling in of forms, commercial transactions, internet searches, use of social network platforms, webmail, loyalty and discount programs - including websites, retail, banks, drug stores and health plans, among many other interactions. Info yield can be carried in either an active way, when the user consciously inputs or allows such capture, or passive, when data is collected without actual acknowledgment - for instance when conversations or images are recorded, when messages are read by algorithms, or GPS history is logged.

The goal of this paper is to recognize the new generation of laws that have been created worldwide, and that have in their cores the establishment of principles, limits, and responsibilities to those that produce and consume data of individuals. Initially, concepts related to the data business as well as a brief history of such industry will be introduced. Next, some of the events related to the leak of personal data and that added to discussing the need for regulations will be illustrated. At last, legislation that was recently enacted in the United States, Europe and Brazil will be acquainted, and their effects on the data industry will be debated.

2   Data Brokers

Data brokers are companies that collect information, including personal information about consumers, from a wide variety of sources for the purpose of reselling such information to their customers for various purposes, including verifying an individual's identity, differentiating records, marketing products, and preventing financial fraud (Federal Trade Commission 2012). For the extraction (and presumption) of knowledge, statistical algorithms are used, and, nowadays, Horvitz and Mulligan (2015) add that machine learning techniques are also used, which can facilitate making the leaps across informational and social contexts, generating inferences.

Big Data, as described by De Mauro et al. (2016), "is the information asset characterized by such a high volume, velocity, and variety to require specific technology and analytical methods for its transformation into value". This definition was picked for this paper for its objectivity and simplicity, but considering the ubiquity of the noun, such is not uncontested, as Ward and Baker (2013) explain: "owing to a shared origin between academia, industry and the media there is no single unified definition, and various stakeholders provide diverse and often contradictory definitions".

The abundant volume of information that people generate every second didn't take long to have its potential recognized, becoming, beyond a commodity, a whole new specialized business. As Sevignani (2013) explains, "commodification is the process of making things exchangeable on markets, either actually and/or discursively by framing things as if they were exchangeable". Roderick (2014) adds that "the growth of companies [...] has facilitated a shift in attention from production-oriented to marketing-oriented strategies, allowing companies to tap into and encourage (ir)rational purchase behavior". More than random or spontaneous data, especially nourished by the massive scale of the social networks, Big Data is also constructed from individuals' data, and that is where its real value resides.

The inception of the data broker industry brought, at its foundation, an extensive list of attracted parties. Regular customers range from banking institutions and financing agencies, seeking risk mitigation and fraud detection, to service providers from niche markets and online retailers, who are interested in consumption profiles, to politicians and candidates, who seek to know - and to influence - their audiences, and also security and intelligence agencies, interested in strengthening their investigations and predictions. Mosco and Wasko (1988) explain the essence of what is happening: "new technology makes it possible to measure and monitor more and more of our electronic communication and information activities. Business and government see this potential as a major instrument to increase profit and control. The result is a pay-per society". Figure 1 illustrates an ordinary flow of information to/from a data broker, describing commonly used sources for capturing data, as well as other public and private actors who participate in the ecosystem.

Although governments usually possess robust databases, eventually they also end up hiring data brokers' services, as according to Stevens (2001), "private companies maintain and organize personal information on individuals in a manner that may not be legally available to government actors". As an example, there is the United States Privacy Act[1], which "establishes a code of fair information practices that governs the collection, maintenance, use, and dissemination of information about individuals that is maintained in systems of records by federal agencies".

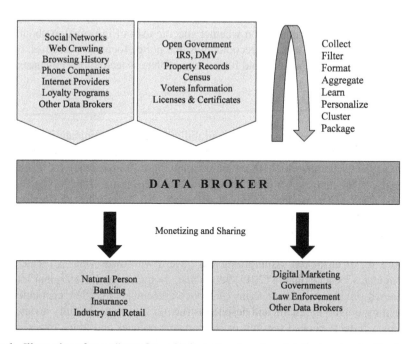

Fig. 1. Illustration of an ordinary flow of information from/to a data broker, inspired by the work of Otto et al. (2007).

---

[1] https://www.justice.gov/opcl/privacy-act-1974.

The comprehension of an individual's decision process through the employment of statistical models allows the establishment of behavioral patterns; additionally, it also grants the development of anticipation on trends (Ostrowski 2013; Ngai et al. 2009). Examples include habit changes, fluctuation on demands, and interest shifts for specific goods. This predictive value complements the intrinsic importance of static data: the greater the collection of entries on someone, the preciser the classification models will be, thus justifying the rush on digital mining.

Knowing information about people makes it possible to cluster them, which means assemble or label them when they share similar characteristics - or according to requirements and attributes pertinent to whoever is interested, from socioeconomic profiles to consuming inclinations. This is a capability inhabited by a latent ethical impasse, as it opens the possibility of the usage of variables that are not only merely demographic, but that may imply questionable context. Features that nowadays are not seen as politically correct involve race, religion, gender, age, and income, among others – and in some legislations could also be a crime. Therefore, linking digital profiles to automated decision-making algorithms may (inadvertently or purposely) lead to discriminatory results, as pointed in the Big Data and Privacy report[2] prepared for The US President's Council of Advisors on Science and Technology in 2014. The report examined the nature of current technologies for managing and analyzing Big Data and for preserving privacy, it considered how those technologies are evolving, and it explained what the technological capabilities and trends imply for the design and enforcement of public policy intended to protect privacy in Big Data contexts. Among its conclusions is the recommendation that policy should focus primarily on whether specific uses of information about people affect privacy adversely. It also recommends that policy focus on outcomes, on the "what" rather than the "how," to avoid becoming obsolete as technology advances.

3 Data Privacy

Although a growing market and virtually multibillionaire, the dissemination of the data broker business model did not occur without questioning. The debates orbit around recurrent subjects, among which, three are commonplace: transparency for when data is captured, loss of control over one's anonymity, and the sharing of the profits.

On the (lack of) transparency subject, Reyman (2013) describes the reality that "terms-of-use policies that describe data collection and use are required by law, but these are lengthy and difficult to understand when read at all" and that data is often obtained on social web technology trade-offs from "tacit agreements that users enter into, and a set of unspoken assumptions that govern who owns what is created and how it circulates." Gangadharan (2017) added that, during research, "marginal Internet users ignored privacy policies or terms of service agreements that they encountered", and "when signing up for email, and despite instructors' advice to carefully review user agreements, students clicked through or past privacy policies and terms of service in order to complete the registration, suggesting these notification mechanisms functioned as meaningless accessories to the new learner's Internet experience".

---

[2] https://obamawhitehouse.archives.gov/sites/default/files/microsites/ostp/PCAST/pcast_big_data_and_privacy_-_may_2014.pdf.

The next common question refers to where do the data end up after all (?), and who has access to it (?), key issues on the argument of anonymity control. In a non-regulated system, information can be sold and distributed without acknowledgments or even accountability of the transactions. In that prospect, Rachels (1985) explains that the value of privacy is "based on the idea that there is a close connection between our ability to control who has access to us and to information about us, and our ability to create and maintain different sorts of social relationships with different people". Roessler and Mokrosinska (2013) add that "the control and regulation of informational privacy should be viewed not only under the perspective of individual rights, but also as being necessary for social interactions themselves, and therefore as relevant to the integration of society".

Another controversy that is consistent concerns the earnings from third-party information: if companies in such business make extraordinary profits with data that are essentially generated by people, where are my paychecks? In that direction, Malgieri and Custers (2018) describe how "personal data of individuals represent monetary value in the data-driven economy and are often considered a counter-performance for 'free' digital services or discounts for online products and services", and point out that "individuals do not seem to be fully aware of the monetary value of their personal data and tend to underestimate their economic power within the data-driven economy and to passively succumb to the propertization of their digital identity".

In a nutshell, the essential aspect of the data broker industry, as emphasized by Crain (2018), is the asymmetrical loss of privacy: "people are opened up to increasingly extensive forms of monitoring, while the institutions doing the monitoring and the information they collect remain hidden from view. [...] Privacy asymmetry as a descriptive category is especially salient for the data broker industry, which has long operated without public awareness or direct regulatory oversight. The privacy of those under watch is undermined, while the watchers themselves operate with substantial freedom from scrutiny".

As technology advanced and the Internet's popularity escalated, there was a great increase in the availability of digital knowledge, as much in the Web as in private databases. Despite the fact that the flourishing of online information might also have dilated the offering of raw material and the number of players that use them, data aggregators have always existed. The catalysts of public objection and the beginning of the model exposure, after decades of progressive exploring, were events known as data breaches, as we will see next.

4   The Exposure of the Data Brokerage Industry

A data security breach occurs when there is a loss or theft of, or other unauthorized access to, sensitive personally identifiable information that could result in the potential compromise of the confidentiality or integrity of data (Stevens 2012). Legislation that addresses such cases usually requires the events to be made public, and both the potentially affected individuals and regulatory agencies to be informed. The obligation to make the facts known is broad, reaching not only data brokers but any private, non-profit or public organization, regardless of their area (health, education, insurance, finance, etc.).

Stevens (2012) explains that security breach notification laws generally follow a similar framework and can be categorized into several standard elements: (1) delineating who must comply with the law; (2) defining the terms "personal information" and "breach of security"; (3) establishing the elements of harm that must occur, if any, for notice to be triggered; (4) adopting requirements for notice; (5) creating exemptions and safe harbors; (6) clarifying preemption and relationships to other federal laws; and (7) creating penalties, enforcement authorities, and remedies.

The significance of the expanding expenses in cybersecurity, with the added intent of also preventing - or minimizing - breaches, can be observed in Fig. 2, where it is illustrated the annual spending, in the United States, in that area.

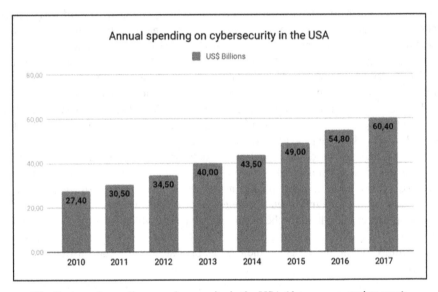

Fig. 2. Annual spending on cybersecurity in the USA (data source: statista.com).

In turn, from the annual records of data breaches in American organizations for the decade 2008 to 2017 (Fig. 3), despite steady expenses in cybersecurity, employee training and expanding regulations concerning individual data maintenance, incidences have been following a stable pattern. Even though the statistics display apparent regularity on the annual number of events illustrated in Fig. 3, on the other hand, the volume of compromised records has been following a rising trend, as can be seen in Fig. 4. It is worth noting that, according to the research, one record corresponds to one data entry, but not necessarily one single individual, as in a computational system, the same person may own several user accounts (for instance, using different email addresses). Considering this ascending scenario, one possible explanation could be the increase in database sizes, proportional to the popularizing of the social networks and the employment of Big Data technologies for capturing information.

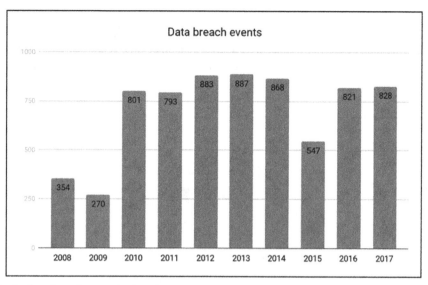

Fig. 3. Data breach events in American organizations in the decade 2008–2017 (data source: pri vacyrights.org).

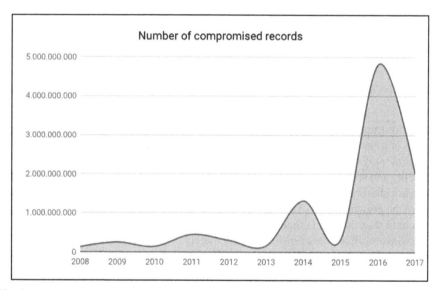

Fig. 4. Progress of the number of compromised records, annually, in data breach events (data source: privacyrights.org).

Data breaches can be categorized according to their causes or origins, which are not limited to cyber-attacks from hackers and malware (although most of the incidents are based on those, granted Fig. 5). They are also considered the cases of unintended disclosure (sensitive information posted publicly, mishandled or sent to the wrong party), physical loss (paper documents or portable devices that are lost, discarded or stolen), insider (someone with legitimate access intentionally breaches information), fraudulent transactions involving debit and credit cards, and finally the unknown cases. Unauthorized access to data causes direct losses due to financial fraud, identity theft, and industrial espionage, besides indirect losses such as reputation and asset depreciation.

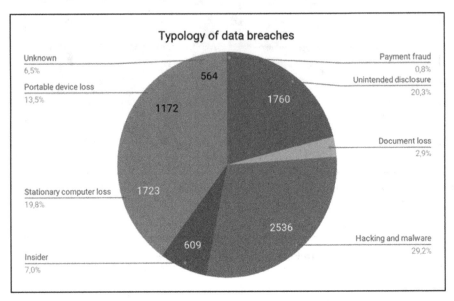

Fig. 5. Typology of data breach events between 2008 e 2017 (data source: privacyrights.org).

Identity fraud, according to Gordon et al. (2007), may be defined as the misuse of personal or financial identifiers for personal gain or to facilitate other criminal activity. Such gains may be obtained from online shopping scams, usage of stolen cards, or controlling and spending over someone else's accounts. Information leaks are often the fuel and/or the beginning of cybercrimes, and the increasing statistics of such frauds, added to the growing cases of data breaches, ultimately brought the attention to a market that up until then operated quietly.

Those who suffer from crimes of identity theft or misuse of personal data are customarily left, often with little or no assistance, with the necessary bureaucracy to reestablish their names and losses. Side with the clamor of the victims, there has been some support from nonprofits and research centers that turned their attention to personal data protection. Among organizations that became distinguished on advocating for more privacy, preeminent cases are the EPIC - Electronic Privacy Information Center[3], and the Privacy International[4], which are both plaintiffs on several civil lawsuits about alleged privacy abuse from tech companies.

---

[3] http://epic.org.

[4] http://privacyinternational.org.

One can adduce that there has been a first era of exploring and preying on personal data on the Web, due to the lack of regulations and auditing. Recent movements, with the enactment of laws specifically designed with the focus on information privacy, point towards a paradigm shift, where there's empowerment to the users while managing their data. In the next section, examples of the legal innovations that sustain this transition will be presented.

5 Personal Data Usage Regulations

It has been observed that technology is likely to evolve faster than the legislative processes. Therefore, the classical narrative applied to the approach of a new problem is to try to address it with existing older laws, up until the creation and establishment of specific regulations. In Brazil, recent examples of this reality are the laws 12.965/2014[5] (known as Marco Civil) and 13.640/2018[6], which regulates the paid transport of passengers by private drivers (like Uber and Cabify, business models created by portable apps). Before the Marco Civil, duties and rights of users and internet providers were disciplined by the Civil Code, Penal Code, Consumer Defense Code and the Constitution (all of them older than the Internet itself).

Pires and Cauchie (2011) state that a legislative change is a deviation chosen by the political system to create an innovation in the law structure, so it can be said, both theoretically and empirically, that it performs an innovative result. Once that innovation is a political operation, but belongs to the legislation (and not to the systems of thought), it produces, at the same time, but in another strand, a special impact on the normal and cognitive arrangements of the law system.

The motif of personal data privacy started to get more attention upon scandals involving information security leaks, obligated to be disclosed by laws that came at the beginning of the third millennium. One example was the enactment of a pioneer Californian bill[7] in 2002, which requires "a state agency, or a person or business that conducts business in California, that owns or licenses computerized data that includes personal information, as defined, to disclose in specified ways, any breach of the security of the data, as defined, to any resident of California whose unencrypted personal information was, or is reasonably believed to have been, acquired by an unauthorized person".

In light of the spreading number of exposure reports (displayed in Fig. 4) and of identity theft crimes, consequently, the model of data acquisition commenced being challenged. The next generation of laws, after a maturing cycle that took ten to fifteen years, is presenting itself much more rigorous when it comes to managing personal data. The main example was the enactment, in 2018, of the European General Data Protection Regulation (GDPR)[8], which affects companies that conduct businesses in the European

---

[5] http://www.planalto.gov.br/ccivil_03/_ato2011-2014/2014/lei/l12965.htm.

[6] http://www.planalto.gov.br/ccivil_03/_ato2015-2018/2018/Lei/L13640.htm.

[7] http://www.leginfo.ca.gov/pub/01-02/bill/sen/sb_1351-1400/sb_1386_bill_20020926_chaptered.pdf.

[8] https://eur-lex.europa.eu/legal-content/EN/TXT/?qid=1528874672298&uri=CELEX%3A32016R0679.

Union, regardless of where they're headquartered. The major changes introduced by it were synthesized in Fig. 6.

Clear language	
Before	After
Often businesses explain their privacy policies in lengthy and complicated terms	Privacy policies will have to be written in a clear, straightforward language
Consent from user	
Before	After
Businesses sometimes assume that the user's silence means to consent to data processing, or they hide a request for consent in long, legalistic, terms and conditions - that nobody reads	The user will need to give affirmative consent before his/her data can be used by a business. Silence is no consent
More transparency	
Before	After
The user might not be informed when his/her data is transferred outside the EU	Businesses will need to clearly inform the user about such transfers
Sometimes businesses collect and process personal data for different purposes than for the reason initially announced without informing the user about it	Businesses will be able to collect and process data only for a well-defined purpose. They will have to inform the user about new purposes for processing
Businesses use algorithms to make decisions about the user based on his/her personal data (e.g. when applying for a loan); the user is often unaware of this	Businesses will have to inform the user whether the decision is automated and give him/her a possibility to contest it

Fig. 6. Key points established by the European General Data Protection Regulation (Source: https://ec.europa.eu/commission/sites/beta-political/files/data-protection-factsheet-changes_en.pdf)

Stronger rights	
Before	**After**
Often businesses do not inform users when there is a data breach, for instance when the data is stolen	Businesses will have to inform users without delay in case of a harmful data breach
Often the user cannot take his/her data from a business and move it to another competing service	The user will be able to move his/her data, for instance to another social media platform
It can be difficult for the user to get a copy of the data businesses keep about him/her	The user will have the right to access and get a copy of his/her data, a business has on him/her
It may be difficult for a user to have his/her data deleted	Users will have a clearly defined "right to be forgotten" (right to erasure), with clear safeguards
Stronger enforcement	
Before	**After**
Data protection authorities have limited means and powers to cooperate	The European Data Protection Board grouping all 28 data protection authorities, will have the powers to provide guidance and interpretation and adopt binding decisions in case several EU countries are concerned by the same case
Authorities have no or limited fines at their disposal in case a business violates the rules	The 28 data protection authorities will have harmonized powers and will be able to impose fines to businesses up to 20 million EUR or 4% of a company's worldwide turnover

Fig. 6. (*continued*)

Brazil also followed the international movement and introduced an update to the Marco Civil law of 2014 by the enactment of law 13.709/2018[9], which addresses personal data protection specifically. In that bill, ten cardinal principles are instituted, in consonance with values that are observed in the European version, and which are described in Fig. 7.

---

[9] http://www.planalto.gov.br/ccivil_03/_Ato2015-2018/2018/Lei/L13709.htm.

Principle	Description
purpose	data must be handled with legitimate, specific and explicit purposes, the user has to be informed, and data should not be processed later on for different reasons than it was initially acquired for
suitability	data must be handled in a way that is compatible with the goals informed to the user
necessity	data handling must be restricted to the minimum necessary to fulfill the purpose it was acquired for
free access	user's access must be provided in a free and facilitated way to information about how and how long their data will be handled
data quality	users must be assured about the correctness, clearness, relevance, and currentness of their data
transparency	users must be given clear, precise and easily accessible information about their data usage and handlers
security	establishment of administrative and technical rules to protect personal data from unauthorized access and from incidental situations such as destruction, loss, alteration, or disclosure
prevention	following of means to prevent damage due to data mishandling
non-discrimination	data cannot be used for abusive, illicit or discriminatory means
accountability	handlers must prove effective actions to obey and enforce such principles

Fig. 7. Principles that must be obeyed on personal data handling (source: Brazilian law 13.709/2018).

That law still establishes, as a requirement for data acquisition and processing, the user's consent, plus his/her right to access, manage, correct, and eliminate the data. Among the penalties for transgressions, there's a fine that can reach up to US$ 12 million. The milestone from the Californian bill that brought up attention to the data exploration business was also not forgotten: any security incident or breach that might lead to risk or damage must be disclosed to the corresponding individuals and the authorities.

Complementing the law 13.709/2018, in July 2019 was enacted the law 13.853/2019, which created the National Data Protection Authority, agency that has technical and decisory autonomy, is bound to the President's Office, and who has the responsibility (amidst others) of watching over personal data protection, overseeing that the rules are followed properly, and applying penalties.

Yet, regardless of Brazilian efforts, it is fair to point that the country was not the first in Latin America to implement those actions; Uruguay had a personal data protection law since 2008 (Ley 18.331[10]), while Argentina has had such legal framework since 2000, based on law 25.326[11]. The importance of those mechanisms was recognized by the European Commission, which regarded both countries "as providing an adequate level of protection for personal data as referred to in Directive 95/46/EC", per decisions 2003/490/EC[12] and 2012/484/EU[13]. Paraguay, in turn, has a bill[14] that was filed in March 2019 ("Proyecto de Ley de Protección de Datos Personales") and is still being discussed on their senate. Over the next section, there will be presented some ideas about adapting the data brokerage business to the new legal scenario described.

6 Proposals and Perspectives

For decades, players known as data brokers operated in a market with little to no regulation, where transactions between corporations (and even governments) were conducted without restrictions and free from public scrutiny. In that context, digitally, pretty much everything was possible: from capturing to buying and selling to sharing information, including that which was mined and statistically inferred (such as the clustering of profiles or the prediction of trends).

In a first moment, laws were crafted to make public those events known as data breaches, where it became mandatory that individuals were notified when their information had been accessed by unauthorized third parties, and which eventually brought attention to a whole market of personal data. The ensuing annual reports, containing a growing number of compromised records - despite the spending on cybersecurity, added to the advocates for privacy, and the tension from organizations that act on behalf of data protection, eventually contributed to the recent advent of a new generation of harsher regulations, already in a global scale.

Modern and important milestones were the European General Data Protection Regulation and, in Brazil, law 13.709, both from 2018. Such novel laws, whose restraints directly affect enterprises in the data brokerage business, suggest a possible downfall to the age where personal information was treated as a commodity. Nonetheless, adaptations to a new paradigm, more focused on privacy and data protection, are feasible, respecting the users and their individual authority on controlling who they are (their digital identities) and what they produce (their generated data).

Values and principles that guided the creation of data protection laws comprise (preceding) consent from the user, transparency, and purpose, premises that must be complied with and also considered while designing the business models. Shifting the spectrum from using third party data as sheer input, and bringing the original sources closer - the users - as partners and suppliers, it is conceivable to depict the continuity of several segments. With a certain level of anonymity or voluntary exposure, products

---

[10] https://www.impo.com.uy/bases/leyes/18331-2008.

[11] http://servicios.infoleg.gob.ar/infolegInternet/anexos/60000-64999/64790/norma.htm.

[12] https://eur-lex.europa.eu/legal-content/EN/TXT/HTML/?uri=CELEX:32003D0490.

[13] https://eur-lex.europa.eu/legal-content/EN/TXT/HTML/?uri=CELEX:32012D0484.

[14] http://silpy.congreso.gov.py/expediente/115707.

such as behavior prediction and consumer profiling, or a wide range of classifiers, might still be appealing and functional.

From a collaborative perspective, new proposals arose, such as the policy framework for user data sharing by Iyilade and Vassileva (2013), based on the idea of a market. In that concept, applications can "offer and negotiate user data sharing with other applications according to an explicit user-editable and negotiable privacy policy that defines the purpose, type of data, retention period and price".

Malgieri and Custers (2018) investigated different models for quantifying the value of personal data, analyzing whether consumers/users should have a right to know the value of their data; the authors also discussed active choice models, in which users are offered the option to pay for online services, either with their personal data or with money. The conclusion, however, was that these models are incompatible with current data protection laws.

Tona et al. (2018) presented "a conceptual design for an artifact that will raise awareness amongst individuals about Big Data ethical issues and help to restore the power balance between individuals and organizations". Their proposal was constructed upon five dimensions, derived from the European GDPR, which are consent, the right to be forgotten, the right to access, data portability and data circulation. All those pillars are arranged over a foundation that would allow several collaborative interactions such as replying, commenting, reviewing, rating and tagging data.

Observing the ubiquity of mobile smartphone usage and the ensuing massive generation of data from those devices - locations, movements, images, video, text, and even health data, which is ordinarily uploaded to content-service providers, Mun et al. (2010) presented a privacy architecture named Personal Data Vaults. In their proposal, individuals retain ownership of their data, which can be reviewed and filtered before being shared, exploring three mechanisms for managing data policies: granular access control lists, trace-audit and a rule recommender, which provides a high-level interface for setting sharing policies.

It can be concluded that, in spite of the strictness of the novel privacy laws, albeit fresh and assigning tech companies to an adjustment cycle, there are several studies and initiatives towards creating tools (frameworks, models and architectures) that invest the users with more power to control their personal information. Such possibilities enforce a pattern shift away from the commoditization of data by the brokers - instead of the lone extinction of their model. Opportunities, therefore, might be in the effective deployment of collaborative platforms and products, which should have their primary focus on the value that has been so emphasized: the transparency.

References

Castells, M.: The Power of Identity, vol. 14. Wiley, New York (2011)

Prins, C.: Property and privacy: European perspectives and the commodification of our identity. Inf. Law Ser. **16**, 223–257 (2006)

Federal Trade Commission: Protecting Consumer Privacy in an Era of Rapid Change: Recommendations for Businesses and Policymakers. Federal Trade Commission, Washington, DC (2012)

Horvitz, E., Mulligan, D.: Data, privacy, and the greater good. Science **349**(6245), 253–255 (2015)

De Mauro, A., Greco, M., Grimaldi, M.: A formal definition of Big Data based on its essential features. Libr. Rev. **65**(3), 122–135 (2016)

Ward, J.S., Barker, A.: Undefined by data: a survey of big data definitions (2013). arXiv preprint arXiv:1309.5821

Sevignani, S.: The commodification of privacy on the Internet. Sci. Public Policy **40**(6), 733–739 (2013)

Roderick, L.: Discipline and power in the digital age: the case of the US consumer data broker industry. Crit. Sociol. **40**(5), 729–746 (2014)

Mosco, V., Wasko, J. (eds.): The Political Economy of Information. University of Wisconsin Press, Madison (1988)

Stevens, G.M.: Data brokers: background and industry overview. Wall Street J. **6**(5), 552a (2001)

Otto, P.N., Antón, A.I., Baumer, D.L.: The ChoicePoint dilemma: how data brokers should handle the privacy of personal information. IEEE Secur. Priv. **5**(5), 15–23 (2007)

Ostrowski, D.A.: Identification of trends in consumer behavior through social media. In: 17th World Multi-conference on Systemics, Cybernetics and Informatics, WMSCI 2013, Orlando, Florida, pp. 9–12, July 2013

Ngai, E.W., Xiu, L., Chau, D.C.: Application of data mining techniques in customer relationship management: a literature review and classification. Expert Syst. Appl. **36**(2), 2592–2602 (2009)

Gordon, G.R., Rebovich, D.D.J., Choo, K.S.: Identity fraud trends and patterns. Center for Identity Management and Information Protection, Utica College (2007)

Reyman, J.: User data on the social web: authorship, agency, and appropriation. Coll. Engl. **75**(5), 513–533 (2013)

Gangadharan, S.P.: The downside of digital inclusion: expectations and experiences of privacy and surveillance among marginal Internet users. New Media Soc. **19**(4), 597–615 (2017)

Rachels, J.: Why privacy is important. In: Ethical Issues in the Use of Computers, pp. 194–200 (1985)

Roessler, B., Mokrosinska, D.: Privacy and social interaction. Philos. Soc. Crit. **39**(8), 771–791 (2013)

Malgieri, G., Custers, B.: Pricing privacy – the right to know the value of your personal data. Comput. Law Secur. Rev. **34**(2), 289–303 (2018)

Crain, M.: The limits of transparency: data brokers and commodification. New Media Soc. **20**(1), 88–104 (2018)

Stevens, G.: Data Security Breach Notification Laws. Congressional Research Service (2012)

Pires, A.P., Cauchie, J.F.: Um caso de inovação "acidental" em matéria de penas: a lei brasileira de drogas. Revista Direito GV **7**(1), 299–329 (2011)

Iyilade, J., Vassileva, J.: A framework for privacy-aware user data trading. In: Carberry, S., Weibelzahl, S., Micarelli, A., Semeraro, G. (eds.) UMAP 2013. LNCS, vol. 7899, pp. 310–317. Springer, Heidelberg (2013). https://doi.org/10.1007/978-3-642-38844-6_28

Tona, O., et al.: Towards ethical big data artifacts: a conceptual design. In: Proceedings of the 51st Hawaii International Conference on System Sciences, January 2018

Mun, M., et al.: Personal data vaults: a locus of control for personal data streams. In: Proceedings of the 6th International Conference, p. 17. ACM, November 2010

Librarianship in the Age of Data Science: Data Librarianship Venn Diagram

Alexandre Ribas Semeler[1]([mail]) [iD] and Adilson Luiz Pinto[2] [iD]

[1] Geosciences Institute, Federal University of Rio Grande do Sul, Porto Alegre, Brazil
alexandre.semeler@ufrgs.br
[2] PGCIN, Federal University of Santa Catarina, Florianópolis, Brazil
adilson.pinto@ufsc.br

Abstract. A Venn diagram is used in mathematics to graphically symbolise properties, axioms, and problems concerning sets and their theories. Thus, this study applied a Venn diagram to describe a theoretical background for data librarianship as a field relating to information science, e-science, and data science. Data librarianship is a new area of study that is located within the thematic core of the triad. The first set on the proposed Venn diagram is information science. Information technology concepts are fundamental to the comprehension of data librarianship in the context of information science. The second set is e-science, an innovative field that incorporates software and hardware that have been built by technology into science. The third set is data science, a way of representing data-driven research in the most diverse knowledge fields; it is a set of the skills, methods, techniques, and technologies of statistics and computer science used to extract knowledge and to create new products and services from data. To ensure greater comprehension of data librarianship, a relatively new field, we suggest some reading materials. The formal discipline of data librarianship is yet to be established in many countries across the globe. Thus, there is the lack of adequate information and certification on data librarianship.

Keywords: Data librarianship · Information science · E-science · Data science

1 Introduction

A Venn diagram is often used in mathematics to graphically symbolise properties, axioms, and problems concerning sets and their theory. The intention of this study is to apply a Venn diagram to describe a background for data librarianship as a practice field in relation to others fields as information science, e-science, and data science.

We propose that data librarianship in nowadays is a practice field located within the thematic core of the triad (information science, e-science, and data science). In this sense, the data landscape manifests as an area of growing research and a new practice that is being adopted by data librarians and other information scientists. Most of the investigations on e-science and data science are oriented towards the practice, usage, and consumption of data. In this sense, this paper proposes a Venn diagram

© ICST Institute for Computer Sciences, Social Informatics and Telecommunications Engineering 2020
Published by Springer Nature Switzerland AG 2020. All Rights Reserved
R. Mugnaini (Ed.): DIONE 2020, LNICST 319, pp. 118–130, 2020.
https://doi.org/10.1007/978-3-030-50072-6_10

to describe data librarianship, to provide a theoretical background for data librarians. Data librarianship is a field that will incorporate information science, e-science, and data science in libraries. Therefore, it is possible to say that the current tendencies in data librarianship research points to subjects related to [...] intensive use of data leads to a technological data information environment that is ideal for studies aimed at models, techniques, and technologies used for data management and curation. This data environment must have competencies related to the creation, management, and preservation of data [1].

Information science is seen as a field of knowledge that involves a series of disciplines, such as computer science and librarianship. In this sense, the object of study of information science is information and the changes caused by the use and re-use of digital technology. Data in digital form are a new form of information generated by all human activities. We believe in a technological continuum that relates with the emergence of digital data in information science and modifies this discipline so that it can investigate the properties of information in all digital contexts in relation to all human activities.

Data librarianship in contexts of data-intensive sciences is related with e-science and data science process. E-science is the term adopted to refer to the use of computational technology during the practice of scientific investigation, which includes preparation, experimenting, data collection, results, storage, digital preservation, and accessibility for all the materials generated during scientific research. It can be said that e-science is preoccupied with the processes of representing information and scientific knowledge through technology that provides data-intensive usage [2].

The prioritization of data technology by scientists is a transformation variable that is common to almost all theories and scientific procedures. Currently, data science is regarded as the theory and practice of extracting knowledge from data; it is concerned with the creation of products and services based on data [3].

Thus, we propose these disciplines as the theoretical pillars for the investigation of data librarianship as a new area of study for librarians. Ultimately, our proposition is a Venn diagram for discussing the theoretical background of data librarianship.

2 Data Librarianship Venn Diagram

We believe that data librarianship is a central filed suitable for practical application to information science, e-science and data science in libraries. The field of data librarianship was made essential in the field of information by the influence of technologies and the direct impact of data-intensive science on the activities of data librarians and researchers, demanding that those professionals acquire skills for usage of all type of data. Thus, there is the question of how to delimit the relationship between information science, e-science, and data science and fashion a new practical and theoretical approach to data librarianship. To answer this question, which depicts the general objective of this paper, we propose a Venn diagram to present and summarise our proposal relating to the theoretical background of data librarianship (see Fig. 1).

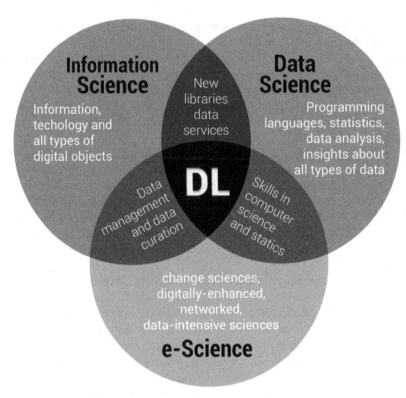

Fig. 1. Data librarianship Venn diagram [4].

The data librarianship Venn diagram (Fig. 1) describes the union and intersection between information science, e-science and data science. With this diagram, it is possible to conceptualise a background of data librarianship. Using three theoretical sets, it is possible to visualise the nucleus of data librarianship as a field: first set (information science) = {information, technology and all types of digital objects}; second set (e-science) = {change sciences, digitally-enhanced, networked, data-intensive sciences}; third set (data Science) = {programming languages, statistics, data analysis, insights into all types of data}.

Next, we present the core of the Venn diagram of data librarianship and the three theoretical sets related to it.

3   The Core of Data Librarianship Venn Diagram

Data librarianship is a area of study that is located within the thematic core of the triad fields, information science, e-science, and data science; its pillars are the new practical and theoretical activities related to data management and curation in digital repositories in the libraries context. The practical approach to data librarianship is geared toward activities such as the collection, manipulation, analysis, and visualisation of data, which are fundamental for data librarians toward offering new data services and products in libraries and other places of information.

Research into data librarianship has its origins in the mid 2000s. They are the result of the effort of North-American, British, and Canadian librarians to create services and products on advertising, consulting, management, preservation, and elaboration of metadata schemas for the effective incorporation of research data in document collections (books, papers, reports, and others). This effort is not something new; it has been discussed since the first applications of data librarianship in the 1960s, particularly in the creation of data services and data archives [1, 4–9].

Data-driven librarianship is not a new branch of librarianship; it is grounded in a diversity of skills already practised by librarians. Thus, traditional practices are reinforced with the objective of being applied to digital datasets. For instance, cataloging and organising all types of bibliographic materials, preservation and curation, user services and reference services, and consultancy and training, are redesigned to be combined with the new practices that encompass data management, data curation and data sharing. However, data librarianship is not only suited as support for the discovery of new skills in librarianship; it is involved with research data generation and preservation, and is concerned with practically all the traditional functions of the librarian trade such as acquisition and collection development, and collection and organisation of documents, cataloging, and user reference services implementation. Currently, the focus of data librarianship is a library services and new approaches for managing and curating digital data from scientific research [1, 4–9].

One of the main characteristics of data librarianship is a core preoccupation with upstream duties [10]; therefore, it cannot only be concerned about published information, but also with potential data sources, being that its objective is to comprehend how the different data types may be capable of generating useful research information [6, 9].

We propose that academic research on data librarianship has focused on fields such as information science, e-science, and data science. Thus, the next section presents the origins of information science and its unfolding in relation to the concepts of information and technology are presented the first set of Venn diagram.

4   Set 1: Information Science

The first set on the data librarianship Venn diagram is information science. Information science is a field of study where interdisciplinary issues circulate. It considers all forms of interaction between people and information [2]. According to [11], information science is a multidisciplinary field of study, involving several forms of knowledge, given coherence by its focus on the central concept of human-recorded information [2].

The concept of information and technology are fundamental to the comprehension of information science in digital context. Thus, [12] propose that information concept be applied in conjunction with documentation fields such as librarianship, among others aspects applied to technological information systems. The author further explained that author's information is a structured set of codified mental and emotional representations (signs and symbols), modeled with/by social interaction, and capable of being recorded on any material medium, and therefore, communicated in an asynchronous and multidirectional way [12].

According to [13–15] investigations about information in information science suggests that the implementation of critical investigations about the conceptual nature of the basic principles of registered information, including the dynamic utilisation of information in the sciences and the application of methodologies and theories to solve problems generated by computer usage.

Information science looks into theoretical and practical issues related to the behavioural characteristics and flows of information when we incorporate technology-mediated hybrid media such as e-science. The ubiquitous technology driving the pervasive behaviour of digital information has changed our existence that has been increasingly mediated by technology. This is not limited to the sciences but encompasses all domains of knowledge. Under this premise, the concept of technology is introduced [2].

When we say technology is a significant part of the contemporary world, we may be referring to our satisfaction by devices that make our lives more comfortable, our enthusiasm in the face of the possibilities guaranteed by computers and internet, or our fear of the increasingly more potent and sophisticated weapons it heralds. Thus, technology presents itself as a multifaceted reality, not only as objects and sets of objects but also as systems, processes, modus operandi, and mentality [16–18].

The most common way to think technology is to regard it as a technical device, a computer or smartphone, hardware, which is a manner of seeing it as something concrete. In the words of [16], technology manifests itself primarily as objects, as material artefacts made by man, whose function depends on a specific materiality. Under this bias, it is necessary to reiterate that technical devices are, in part, user dependent and using and/or handling them require special skills and training [16–18].

Nowadays, technology can be defined as hardware, as rules, and as systems. First, technology is usually thought of as being comprised of machines or tools—hardware—this is a way of looking at it as something concrete. However, a distinction between these concepts should be made. While users directly manipulate tools, machines are more independent of the user and require skills and training to be used or handled. A second standpoint views technology as rules. Software can be seen as a metaphor for this approach; it involves patterns of meaning and relationships, such as political rules, laws, and scientific principles that are systematically developed. The third way of viewing technology is as systems. This suggests that hardware artefacts and software are technologies, but that they need to be considered in the context of their user [1, 2].

Technology can be a specific form of knowledge, comprising of specific ways of knowing the material world, ways that incorporate scientific knowledge, but also equally possess their own characteristics. In this sense, technology defines itself as a multifaceted, artifact manufacture and usage; a form of human knowledge addressed to create a reality according to our own purposes; know-how; practical implementations of intelligence; humanity at work; nature available to man as a resource; the field of knowledge related to artifact design and the planning of its achievement, operation, adjustment, maintenance, monitoring, under the light of scientific knowledge; Modernity own lifestyle; the totality of methods that can be reached rationally; in all fields of human activity; the material structure of Modernity, in four dimensions: as objects; way of knowledge; specific form of activity or will (determinate human action when facing reality) [16–18].

This examination aims to discuss information and technology to examine the impact of technology on scientific knowledge and information concepts, i.e. research activities, the discourse about scientific culture, and its influence on the processes of interaction between information and technology. When information is mediated by technology, it can reveal the complexity of new ways of scientific investigation [2]. Thus, the next section is a presentation on e-science, the second set of data librarianship Venn diagram.

5 Set 2: E-science as a Contemporary Data and Information Phenomenon

The investigation methods based on the use of hardware infrastructure and scientific software, have allowed the emergence of e-science. In the 1990s, John Taylor, the Director General of the United Kingdom's Office of Science and Technology, created this term to refer to the use of technology to conduct scientific investigations. In United States a 2007 lecture, Jim Gray described the e-science as a new way of doing science. He believes that e-science is the fourth paradigm of science [2, 19–21].

E-science is one of many terms used to describe recent transformations in the scientific enterprise. It is a way to translate methods and scientific disciplines to computers, allowing scientific collaboration on a global scale, in addition to intensive data usage [21].

Thus, it is perceived that, beyond the concern with immense data volume, generally, e-science basic characteristics are being digitally-enhanced, networked, and data-driven. The digitally-enhanced aspect of e-science embodies how e-scientists analyse, manage, gain access to, and share data digitally. Without the technological advances of our time, the amount of data we have access to and are able to create would significantly decrease. Almost exclusively through online databases, networks, shared digital repositories, and digital files, datasets are shared, and therefore, are digitally accessible. Networking is a very important feature of e-science because this is what makes e-science particularly fascinating, the networked and multidisciplinary nature of the field that creates initially unintended links among datasets. E-scientists utilise networked data and materials to formulate new information through cross-comparison and manipulation. This creates another form of scholarly communication that is networked and constructed extemporaneously, depending on the current usage; such as how articles are connected through citations and websites. Therefore, e-science thrives when datasets are shared and accessible. The final characteristic of e-science is that it is data-driven. E-science exploration is achieved through data manipulation; thus, datasets are used as the primary form of experimentation. By manipulating and cross-comparing datasets, researchers are able to find patterns and develop new inquiries across disciplines [22].

Current studies about data-driven sciences amplify themselves when mediated by computers, mainly if networks like the web are included. This gives rise to different methods applicable to data collection and/or usage; and thus, a new way to practise science, based on technological instruments and technical devices created by contemporary computing that has as its basis the data-intensive usage of the world of science. Data-driven sciences opened a new dimension to scientists, the data universe, causing a revolution in the process of scientific thinking.

The e-science phenomenon is responsible for the generation and use of data in large quantities. In the Big Data era, data, information and technology are pervasive. E-science is a contemporary data/information phenomenon. E-scientists believe that data-driven sciences will yield optimum results with the intensive use of technological instruments and devices. The theoretical understanding of these studies is based on appreciating the fact that the nature of scientific things depends on the materiality of technical devices. Therefore, scientific instruments are the key to contemporary scientific practice. This conclusion makes science less about knowledge than about a practice that involves using machines and technical devices [2, 23].

The author [20] introduced the idea of a fourth paradigm in a lecture delivered in California in 2007 for the National Research Council of the United States. He considers computational science as the third type of Science, aiding the various domains of knowledge, for example, the hard sciences such as Astronomy, Physics, and Medicine, amongst other categories of science. He amplifies the idea of e-science, thus, making the term synonymous to the fourth paradigm of science. It is believed that digital technology revolutionised the scientific methods, as posited by computer sciences made new scientific paradigms emerge which modified the way scientific research is.

Investigations in the fields of e-science were simulated against the backdrop of hard science disciplines such as astrophysics, genetics, oceanography, climatology, nuclear physics because the main concern of these fields is the immense volume of data generated during scientific research. E-science channels science transformation towards the intensive usage of huge data volumes.

E-science methods were first used by Biology, Physics and Astronomy researchers in projects such as the GenBank [24], the Large Hadron Collider (LHC), the Panoramic Survey Telescope and Rapid Response System (Pan-STARRS) [25] and other projects in the world of hard sciences. For example, the World Wide Web, invented by Tim Berners-Lee at the European Organization for Nuclear Research (CERN), was created for sharing document and data amongst researchers. For example, the Pan-STARRS project will retain 2.5 petabytes of data each year, when it starts operation [31]. CERN's Great Hadrons Collider (GHC) will generate 50 to 100 PB of data each year, with circa 20 PB stored and processed in a global federation of national arrays linking 100 thousand CPUs. Therefore, scientists and institutions need a model and better set of practices to provide balanced hardware architecture to the corresponding software capable of dealing with those data volumes [21, 24, 25].

The large data volume generated by e-science gives rise to data science. Data science is regarded as a field of study connected to data analysis and management. In the next section, we present data science as a scientific theory and practice used for the comprehension of the diverse dimensions of data-intensive sciences.

6 Set 3: Data Science

Data science is a field of knowledge that merges techniques of computer science and statistics. It can be considered as a new area of study for librarians seeking to engage with issues related to data management and analysis. It is a field of knowledge that demands specific expertise, usually related to mathematics and programming languages [1].

The term data science has its origin in disciplines such as Econometrics, Physics, Bio-statistics, computer science, and statistics and engineering. It can be a synonym or related term for concepts such as business analytics, operational research, competitive intelligence, data modelling and analysis, and the discovery of knowledge at databases. The time-line (see Fig. 2) presents data science evolution.

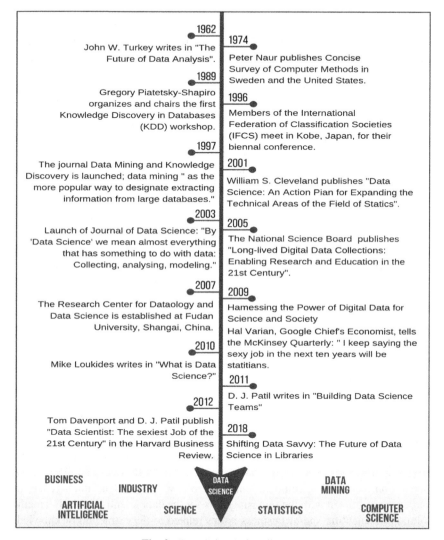

Fig. 2. Data Science time-line [4].

Although data science is often viewed as a new term, it has been used in the field of computer science since the 1960s. [27, 28] reveal that [29] proposed that computer science be called "Datalogy". In the 1990s, the Committee on Data of the International Council for Science (CODATA) established as priority the realisation of studies about advances in data science, CODATA using the term data science as a way of representing data-driven research in the most diverse knowledge fields. In the 2000s, the first journals appeared, such as the Journal of Data Science from the Columbia University in the USA. Finally, data science appears, in current usage, as a combination of data analysis development and the elaboration of new intelligent algorithms in artificial intelligence, on a par with statistics and computer science [26–29].

Data science possesses two basic functions: the analysis and invention of new techniques that may generate insights during the usage of data, which was not possible before the 2000s. In that sense, data science can be seen as the theory and practice of extracting knowledge from data. Being a scientific discipline, which is preoccupied with creating products and services from data, its aims are to use data-driven sciences for the benefit of the environment, such as the use of Internet products as data repositories, climatic sensors, GPS, and the most recent tendency to connect objects to network [27, 28].

However, one of the main investigation problems in data science is geared towards establishing it as a science of data and not a mere extension of statistics and computer science techniques. Data science aims to analyse and understand contemporary phenomena using data. [30]. The data science is as a new area of study for information professionals seeking to engage in data management and data analysis issues based on statistics and computer science [27, 28].

In a recent paper, [31] presented a definition of data science and the crucial role of librarians in its development. Advances in statistics and computer science, combined with an abundance of data, have given rise to a new professional ecosystem. This ecosystem is concerned with generating insight from data to inform decision-making. Data science methods and products have transformed commerce, health care, and government, and they will continue to transform other sectors [1].

Data science requires knowledge and techniques to data analysis, including statistics and computer programming, requires systematic thinking combined with a creative approach to solve general pragmatic problems. Data science combines data mining, machine learning, system design, distributed computing, statistics, industrial engineering, domain knowledge, visualisation, and Big Data [32, 33].

The scientific status of data science is connected to different objectives such as: (a) the study of scientific data, which sees data science as a method and technology used to conduct scientific researches through scientific data management; it refers to the data used during scientific researches; (b) the study of business data, which sees data science as a technique for creating data products or services, and as a methodology to generate solutions for problems in a business environment (business intelligence); (c) the integration of statistics, computational technology, and artificial intelligence; this objective is preoccupied with the requisite skills for the data scientist; and (d) a method for solution of scientific and/or business issues related to the extraction of knowledge from data; that is, an objective which integrates all other objectives [1, 32, 33].

Therefore, following the identification of the combination of disciplines and the practical development process that compose data science, it becomes imperative to identify the skills needed by data scientists. Data science can be regarded as a new area of study for data librarians seeking to become involved in issues related to data management and analysis. Thus, data librarians should understand the complexity and variation associated with data science praxis because it provides new methods and practices for data librarianship. A data librarian needs not become a programmer, statistician, or database manager, but should possess foundational competencies in the languages and programming logic of computers, databases, and information retrieval tools [1].

Finally, we will suggest some readings, courses and suggested themes for investigation, in relation to the triad, information science, e-science, and data science.

7 Final Remarks: Literature, Courses, and Future Work in Data Librarianship

The final question is related to the difficult localisation of literature about data librarianship. It reveals that data librarianship is still developing and needs attention from the professionals of traditional librarianship. In that sense, in order to acquire minimum theoretical knowledge in data librarianship, few essential works are recommended. The work, organised by Lynda Kellam and Kristi Thompson in 2016, is the anthology of articles, *Databrarianship: the academic data librarian in theory and practice*, which presents diverse questions about data librarianship and research data usage based on the most distinct scientific disciplines [8]. Furthermore, another significant work The book *Data Librarian's Handbook*, written in 2016 by Robin Rice and John Southall; it focuses on the data librarian's field of action [5, 34, 35].

Examining the questions of e-science, the book writings by Tony Hey; Stewart *T*ansley and Kristin Tolle, *Fourth Paradigm: Data-intensive Scientific Discovery* (2009) is another significant reading material. The importance of this text lies in the fact that it sees e-science in all its aspects, in relation to scientific and technological advances, scientific communication or the intensive use of data extracted from technological sensors [35].

The questions related to data science and the requisite skills for the data scientist appear in first-time in the text, *Data Scientist: The Sexiest Job of the 21st Century* published by the Harvard Business Review, in 2012, written by Thomas Davenport and D. J. Patil [36]. Another essential text on data science is *The Data Science Venn diagram*, elaborated by Drew Conway in 2010 available free in the web. The report about data science in libraries was published in 2018, *Shifting to data savvy: The future of data science in libraries*; it presents the synthesis of the discussions on data science as applied in North American libraries. The report represents data science projects in libraries in the USA. The project explores the challenges associated with the implementation of data science in different libraries environments. The report is oriented around four distinct facets of a multi-faceted framework: structures, skills, services, and stakeholders. The report discusses specific perspectives in relation to the lack of skills militating against the practice of data librarianship, in other words, what is missing in librarians' information management skills; and also in relation to data management, in other words, the capability of library managers to understand and appreciate the benefits of the new skills brought by data science to offer organizational and managerial support to libraries [37].

In this sense, to acquire skills, there are some courses and formation courses offered by international universities, whether in the form of proximate or distance learning. For example [38], we have the Research Data Management and Sharing course, offered online for free by the Curating Research Assets and Data Using Lifecycle Education (CRADLE), of the North Carolina University, at Chapel Hill, in partnership with project EDINA, of the Edinburgh University. An online course is the Research Data Management Training MANTRA [39], offered by project EDINA, of the Edinburgh University, focused on research data management. The course is useful for both data librarians and other researchers. These courses offer an introduction to research data management and sharing, life cycle, data management, and best practices in relation to working with data (including organisation, documenting, storage, and security), and research data repositories.

The courses offered by the Library Carpentry initiative website is made up of software and data skills training originally based on lessons developed by the Carpentries. The aim of this website is help library professionals work more efficiently, and potentially teach the skills they have learned to colleagues, students, and researchers. Another course is offered by the Harvard-Smithsonian University, at the Center for Astrophysics, John G. Wolbach, and by the Harvard Library focused on preparing data librarians for acquiring knowledge about data science [40].

To learn statistic techniques and programming languages used in data science, data librarians might opt for courses offered by universities in the field of statistics or computer science. They might opt for completely different formation courses offered by online training platforms such as Coursera, Udemy, Data Science Academy, and EDX.

For further studies, it is suggested that data librarians investigate themes related to research data usage in diverse contexts such as Big Data, Internet of Things, artificial intelligence, informational patterns, and scientific communication. It would be interesting to accomplish a study about the possible analysis of the metadata of these fields and the content stored at research data repositories. Thus, in conclusion, this paper recommends that data librarians should work with research data repositories, supporting researchers, from the first steps of scientific investigations, assisting with the documentation process and providing the assurance that the research data will be preserved, usable, and reusable in the long term. A data librarian must use the fundamental values, ethical principles, skills, and professional knowledge of data librarianship to work with research data, especially at digital repositories.

Finally, data librarians should be able to combine the skills and knowledge related to information science, e-science, and data science, to consolidate the Venn diagram of data librarianship. This diagram was proposed to enable for data librarians to consolidate their practical activities supported based on consolidated theoretical fields such as information science, in addition to new knowledge in fields such as e-science and data science.

Funding. This study was financed in part by the Coordenação de Aperfeiçoamento de Pessoal de Nível Superior - Brasil (CAPES) - Finance Code 001.

References

1. Semeler, A., Pinto, A., Rozados, H.: Data science in data librarianship: core competencies of a data librarian. J. Librariansh. Inf. Sci. **51**(3), 771–780 (2019)
2. Semeler, A., Pinto, A.: E-science: an epistemological analysis based on the philosophy of technology. IFLA J. **43**(2), 198–209 (2017)

3. Nielsen, L., Bulingame, N.: A Simple Introduction to Data Science. New Street Communications, Wickford (2012)

4. Semeler, A., Pinto, A.: Data Librarianship Venn Diagram Handbook (Beta Version). IGEO/UFRGS, Porto Alegre (2019)

5. Rice, R., Southall, S.: The Data Librarian's Handbook. Facet Publishing, London (2016)

6. Koltay, T.: Data literacy for researchers and data librarians. J. Librariansh. Inf. Sci. **49**(1), 3–14 (2017)

7. Tenopir, C., Talja, S., Horstmann, W.: Research data services in European academic research libraries. LIBER Q. 2017 **27**(1), 23–44 (2015)

8. Kellam, L., Thompson, K.: Introduction to Databrarianship: The Academic Data Librarian in Theory and Practice. Association of College and Research Library, Chicago (2016)

9. Xia, J., Wang, M.: Competencies and responsibilities of social science data librarians: an analysis of job descriptions. Coll. Res. Libr. 2014 **75**(3), 362–388 (2015)

10. Gold, A.: Cyberinfrastructure, data, and libraries, part 1: a cyberinfrastructure primer for librarians. D-Lib Magazine, 13(9/10) (2007)

11. Bates, M.: The invisible substrate of information science. J. Am. Soc. Inf. Sci. **50**(12), 1043–1051 (1999)

12. Silva, A.M., Ribeiro, F.: Information science and philosophy of information: approaches and differences. In: Demir, H. (ed.) Luciano Floridi's Philosophy of Technology: Critical Reflections. Springer Science+Business Media, Dordrecht (2012)

13. Floridi, L.: Open problems in the philosophy of information. Metaphilosophy **35**(4), 554–582 (2004)

14. Floridi, L.: The Philosophy of Information. Oxford University Press, Oxford (2010)

15. Floridi, L.: Steps forward in the philosophy of information. Etica Politica/Ethics Politics **14**(1), 304–310 (2012)

16. Cupani A.: Filosofia da Tecnologia: um convite. Ed. da UFSC, Florianópolis (2013)

17. Scharff, R., Dusek, V.: Philosophy of Technology: The Technological Condition: Ananthology. Blackwell Publishing, Malden (2006)

18. Kline, J.: What is technology? In: Scharff, R., Dusek, V. (eds.) Philosophy of Technology: The Technological Condition an Anthology. Blackwell Publishing, Malden (2006)

19. Yang, X., Wang, L., Laszewski, G.: Recent research advances in e-science. Cluster Comput. **12**(4), 353–356 (2009)

20. Gray, J.: Jim Gray on eScience: a transformed scientific method. In: Hey, T., Tansley, J., Tolle, E. (eds.) Fourth Paradigm: Data-intensive scientific discovery. Microsoft, Redmond (2009)

21. Jankowski, N.: Exploring e-science: an introduction. Special theme: e-science. J. Comput.-Med. Commun. **12**(2), 549–562 (2007)

22. Shcmillen, H.: Library and Information Science Education and eScience: Current State of ALA Accredited MLS/MLIS Programs in preparing Librarians and Information Professionals for eScienceNeeds. Capstone Projects, Denver (2015)

23. Davenport, T.: Big Data at Work: Dispelling the Myths, Uncovering the Opportunities. Harvard Business Review Press, Boston (2014)

24. Genbank Homepage (2018). https://www.ncbi.nlm.nih.gov/genbank/. Accessed 15 Aug 2019

25. PAN-STARRS Homepage. https://panstarrs.stsci.edu/. Accessed 15 Aug 2019

26. Fay, D.: Earth and environment introduction. In: Hey, T., Tansley, J., Tolle, E. (eds.) Fourth Paradigm Data-intensive scientific discovery. Microsoft, Redmond (2009)

27. Zhu, Y., Xiong, Y.: Towards data science. Data Sci. J. **14** (2015)

28. Chen, L.M.: Introduction: data science and bigdata computing. Mathematical Problems in Data Science, pp. 3–15. Springer, Cham (2015). https://doi.org/10.1007/978-3-319-251 27-1_1

29. Nauer, P.: https://en.wikipedia.org/wiki/Peter_Naur. Accessed 15 Aug 2019

30. Hayashi, C.: What is data science? Fundamental concepts and a heuristic example. In: Hayashi, C., Yajima, K., Bock, H.H., Ohsumi, N., Tanaka, Y., Baba, Y. (eds.) Data Science Classification, and Related Methods. STUDIES CLASS. Springer, Tokyo (1998). https://doi.org/10.1007/978-4-431-65950-1_3

31. Burton, M.: Data science in libraries. Bull. Assoc. Inf. Sci. Technol. **43**(1), 33–35 (2017)

32. Aalst, W.M.P.: Data scientist: the engineer of the future. In: Mertins, K., Bénaben, F., Poler, R., Bourrières, J.-P. (eds.) Enterprise Interoperability VI. PIC, vol. 7, pp. 13–26. Springer, Cham (2014). https://doi.org/10.1007/978-3-319-04948-9_2

33. Voulgaris, Z.: Data Scientist: The Definitive Guide to Becoming a Data Scientist. Technics Publications, Basking Ridge (2014)

34. Koos, J.: The data librarian's handbook, by Robin Rice and John Southall. Med. Ref. Serv. Q. **37**(2), 217–218 (2018). https://doi.org/10.1080/02763869.2018.1439231

35. Hey, T., Hey, J.: E-science and its implications for the library community. Libr. Hi-Tech **24**(4), 515–528 (2006)

36. Davenport, T., Patil, D.: Data scientist: The Sexiest Job of the 21st Century. Harvard Business Review, Cambridge (2012)

37. Burton, M., Lyon, L., Erdmann, C., Tijerina, B.: Shifting to data savvy: the future of data science in libraries. Project Report. University of Pittsburgh, Pittsburgh, PA (2018)

38. Cradle Homepage. https://pt.coursera.org/learn/data-management. Accessed 15 Aug 2019

39. Mantra Homepage. https://mantra.edina.ac.uk/. Accessed 15 Aug 2019

40. Library Carpentry Homepage. https://librarycarpentry.org/. Accessed 15 Aug 2019

Use and Analysis of Network Information

Examining the Linking Patterns and Link Building Strategies of Mainstream and Partisan Online News Media in Central Europe

Andrea Hrckova$^{(\boxtimes)}$, Robert Moro , Ivan Srba , and Maria Bielikova

Faculty of Informatics and Information Technologies,
Slovak University of Technology in Bratislava, Bratislava, Slovakia
{andrea.hrckova,robert.moro,ivan.srba,maria.bielikova}@stuba.sk

Abstract. The presence of external links or sources in the articles are considered as one of the indicators for assessing their quality by a librarian and information community. In this article, we explore linking patterns of the most popular traditional and "alternative" (partisan) digital news media in two V4 countries of Central Europe: Czech and Slovak Republic. Alternative/partisan news media are understood as media that protest against traditional or mainstream media. Fake news as articles containing disinformation can appear in traditional media (e.g. in tabloids) as well as in alternative media. Eighteen most popular mainstream news media and fifteen most popular partisan media in Slovakia and Czech Republic were selected for quantitative and qualitative analysis of links. With this method, more than 171 million of unique domains of hyperlinks from and to the selected online media were collected and analyzed. The argument to conduct this type of research is that partisan news media are gaining popularity in the countries of Central Europe and the linking patterns of these media were rarely examined or compared with traditional media. Quantitative analysis and visualization of hyperlinks was performed using two software systems: Ahrefs and Gephi. We concluded that there are some differences between the linking patterns of mainstream and partisan digital news media that need the research attention not only to follow the communication patterns of digital media, but also to be able to detect the type of the news media automatically.

Keywords: Fake news · Online news media · Alternative media · Partisan media · Mainstream media · Traditional media · Link building · Hyperlink network analysis

1 Introduction to Partisan and Mainstream Online News Media

Fake news, as species of disinformation [5], are understood as news articles that are intentionally and verifiably false and could mislead readers [1]. This definition

© ICST Institute for Computer Sciences, Social Informatics and Telecommunications Engineering 2020
Published by Springer Nature Switzerland AG 2020. All Rights Reserved
R. Mugnaini (Ed.): DIONE 2020, LNICST 319, pp. 133–146, 2020.
https://doi.org/10.1007/978-3-030-50072-6_11

is rudimentary in theory, but the borders of truth in the news media are more blurred in practice. Some online media combine truth with lies, somehow manipulated or one sided opinions, whether it is yellow journalism or biased news with clickbait articles or even real fake news with manipulated images (photoshopping) or videos (deep fake), portals with PR articles or content farms [6].

Therewith, the traditional (mainstream) or in other words established online media are also being attacked by some newcomers to the media scene as "hysterical media that annoy the public by analyzing "lies" as a phenomenon of the times" [15] that behave the same or worse than "conspiracy media", referring mainly to Chomsky and Herman [4]. According to these portals, the goal of these mainstream "truthloving media" is to "indoctrinate the educated and fool the masses" [8]. Some of these portals use an implicit language, stating that "nobody will dictate them, what they write" (Parlamentne listy). To categorize these portals for the purpose of this article, we utilize the term "alternative" or "partisan" media, as they defined themselves against the mainstream (or traditional/established) media.

The online mainstream (or traditional) media are for the purpose of this article defined as media news portals that usually have all the attributes of credible news media: the real name of author and publisher, listed on each website and employing journalists doing fieldwork, whether it is tabloid or serious news. It is important to note that in different countries, the concept of "alternative" and "traditional" media may differ. In Slovakia and Czech Republic alternative media are mostly the media, marked as fake or conspiracy news by some fact checkers (as e.g. konspiratori.sk) that gives them a negative connotation. These media are also connected by their pro-Russian attitude as an "alternative" to the pro-western traditional media. In Hungary, the traditional and established media are mostly owned by pro-government bodies and the real alternatives to these media have a weak position in the news media landscape [9]. In Poland many pro-government right wing media claim to be "alternative" and for the opposition's supporters, completely other online news media are considered "alternative" [13]. Therefore, the term "partisan media" is preferred in this article.

Nevertheless, the focus of this article is not to analyze the concept of truth or the pro- or anti-government attitude of these media. The objective of this article is to analyze the external hyperlinks and linking patterns between the so called mainstream (traditional) and partisan (alternative) media in Czech and Slovak republic and thus assist with the detection of linking patterns as one of the indicators to automatically determine the type of these media. With this objective in mind, these research questions were set: What is the position of alternative/partisan news media in the online news media landscape in Slovak and Czech Republic? Are the (crawlable) linking patterns different in traditional and partisan online news media? Who are the main actors in the networks?

2   Research Antecedents

As a methodology to study hyperlinks among websites, Jackson [7] first suggested that the social network analysis (SNA) methods are applicable. SNA is a research method for identifying structures or patterns between social entities in various social systems based on the relations among the system components (or nodes) rather than on the attributes of individual actors [12]. Hyperlinks are technological capabilities that enable websites or web pages to link with another, but also to exchange information, and maintain cooperation between actors with common background, interest, or projects [10]. A website therefore functions as a node that passes messages from one actor to another and patterns of hyperlinks reflect the communicative choices of the owner [7]. Likewise, external hyperlinks to the websites are set as the indicators of web content quality by search engines like Google, similarly as the citations in the documents. As soon as the number and quality of hyperlinks started to be a part of evaluation of websites by search engine algorithms, the practices of link building as a strategy of creating links on websites within the search engine optimization methods have been widespread. Moreover, the citation analysis is a widespread method for analyzing scholarly articles to determine the cooperation between scholars, therefore hyperlink network analysis (HNA) method might enable the analysis of cooperation on internet websites.

Park [10] mentions four areas of research, where hyperlink network analysis can be utilized: e-commerce and the communication on international, interpersonal and interorganizational level. Park and Thelwal [11] then analyzed hyperlink patterns and social structure on politicians' websites in South Korea. Other authors that utilized HNA for analyzing media are Szabo and Bene [14]. The research objective of these authors was to analyze the position of radical right mass communication channels in the media sphere in Hungarian media network. Nevertheless, HNA, despite its long tradition, remains underutilized in terms of media communication and SNA in social networks is a more common method to analyze data also in the field of fake news and rumors detection. With this article, we show that HNA still reveals important information about the media relationships and linking patterns of media – both traditional and partisan.

The analysis of web hyperlinks structure is also considered as part of webometric research. Webometrics is "the study of the quantitative aspects of the construction and use of information resources, structures and technologies on the web drawing on bibliometric and informetric approaches" [2]. However, in the field of untrustworthy sources, webometrics as a method is associated especially with the analysis of hyperlinks in scientific resources, as e.g. in the article of Bowler, He and Hong [3] that were researching the hyperlinks between websites and blogs in medical domain.

3   Methodology

Two software systems were utilized to obtain and analyze the data: Ahrefs[1] and Gephi[2]. Ahrefs is a crawler, the original purpose of which is to serve search marketing professionals. It collects, processes and stores data, consisting of hyper-

---

[1] https://ahrefs.com/.
[2] https://gephi.org/.

links, keywords and user behavior. Using commercially available crawlers to discover hyperlinks between websites has been a common practice in webometrics [2]. AhrefsBot is also mentioned as the second most active crawler on the internet after Googlebot [16]. Gephi is an open source software for visualization and exploration of graphs and networks, applicable also for link analysis. Thanks to Gephi visualizations, the networks of directed hyperlinks from and to the online news media and the importance of some nodes, based on the automatic calculation of the sum of hyperlinks will be visible.

Data from the most popular Czech and Slovak online news media were obtained throughout July and August 2019 with Ahrefs tool. The popularity of the mainstream news media was identified by available statistics (e.g. Aim monitor) and the popularity of partisan media, where official statistics are often missing, the number of fans on Facebook or the number of hyperlinks was selected as a presumption of their popularity. We selected seven websites of Slovak partisan web media (Dolezite, Hlavne spravy, Infovojna, Parlamentne listy, Protiprud, Slobodny vysielac, Zem a vek) and nine websites of Slovak mainstream media (Aktuality, Cas, Dennik N, Hnonline, Pluska, Pravda, Sme, Topky, Webnoviny). In Czech Republic, where the news media scene is richer, ten websites of Czech partisan media (Ac24, Aeronet, Cesko aktualne, Cz.sputniknews, Parlamentní listy, Pravy prostor, Protiproud, Svobodne noviny, Svobodny vysilac, Zvedavec) and nine Czech mainstream web media (Aha, Blesk, Ct24, iDnes, iHned, Lidovky, Pravo, Reflex and Respekt) were selected. The complete list of the selected media with their description can be found in the appendix. The data that was of our particular interest was the number and domains of incoming and outgoing hyperlinks (robots do follow and also no follow) to and from the selected media.

These data were downloaded in .csv format and manually processed for the needs of Gephi in September 2019 (Table 1, Table 2). To define the weight of edges and the modularity class of the nodes, the number of all hyperlinks (with dofollow and nofollow attributes) was selected. The number of dofollow links was set as an additional attribute to analyze the data, because many of the analyzed hyperlinks had a nofollow attribute. Hyperlinks with nofollow attribute are not officially followed by the search engines and include hyperlinks from discussions or blogs. Hyperlinks with dofollow attribute are commonly considered to pass "weight" and help a website or URL to rank better in organic search results.

Data was also cleaned from hyperlinks to social networks, Google tag manager and hyperlinks created by link shorteners (like bitly). The duplicates were also merged and the sum of the hyperlinks in duplicates was calculated automatically by Gephi. Finally, we collected and analyzed more than 171 million of unique domains of hyperlinks from and to the selected online media (more than 160 million of hyperlinks from and to the mainstream media and more than 11 million of hyperlinks from and to the partisan media). The number of hyperlinks to and from mainstream news media is noticeable higher than the number of links of partisan news media as mainstream media are more established in the news landscape and one of their link building strategies is buying smaller and specialized portals with online history and hyperlinks.

To visualize the data in the graph properly, Fruchtman Reingold layout was selected. To filter the data, degree range $n > 1$ and edge weight of $n > 0.01$ of the total number of hyperlinks (total edge weight) was invoked. The graphs in the article are created from both dofollow and nofollow attributes.

Table 1. Structure of nodes prepared for the visualization in Gephi. Modularity class is the number of all hyperlinks with both dofollow and nofollow attributes

ID	Label	Modularity class	Dofollow
1221022	auto.cz	20874449	20874443
1221023	blesk.cz	7558503	7510575
1221024	reflex.cz	2803439	2803439

Table 2. Structure of edges prepared for the visualization in Gephi. Weight is computed the same as the modularity class, i.e., as the number of all hyperlinks with both dofollow and nofollow attributes.

ID	Source	Target	Weight	Dofollow
1221022	1221022	1222049	20874449	20874443
1221023	1221023	1222049	7558503	7510575
1221024	1221024	1222049	2803439	2803439

4   Results

4.1   What Is the Position of Alternative/Partisan News Media in the Online News Media Landscape in Slovak and Czech Republic?

The position of partisan media in the local news media landscape can be inferred from the graph of all hyperlinks from and to the most popular mainstream and partisan media of the researched country. The graph suggests that there are separated traditional media clusters with more than one million hyperlinks. In this graph (Fig. 1), one popular partisan medium (Hlavne spravy) can be spotted that is co-linked with a traditional medium by a questionable domain. When analyzing this traditional medium (Pravda.sk), its new owner is also the owner of some partisan media. Therefore, we assume that there is one partisan medium (Hlavne spravy) that has a weak position in the online media landscape in our country.

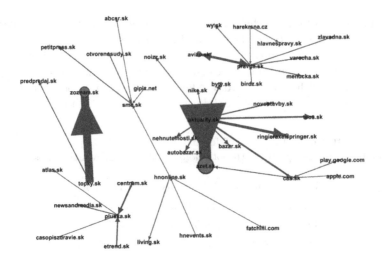

Fig. 1. Network of directed hyperlinks from and to the most popular Slovak partisan and traditional media

According to the analysis in Ahrefs tool, Slovak traditional media rarely link to partisan media, except of tens of thousands of outgoing nofollow links from the mainstream media blogs. We assume that guest blogging is one of the popular link building practice of Slovak partisan media. Although these links are marked as nofollow for the search engines, they are visible for the users and thus connecting the partisan media scene with the mainstream media. Pravda.sk is the only Slovak traditional medium that links heavily with dofollow links to partisan media – more than nine thousand to the most popular alternative medium Hlavne spravy and almost seven thousand to Sputnik news.

The situation is different in Czech Republic, where three big clusters of traditional and partisan news media are visible (Fig. 2). Half of one cluster on the right side belongs to the partisan news media (ceskoaktualne, az24, zvedavec and parlamentnilisty), all linked with a questionable portal harekrsna.cz. This portal also links these news media to the mainstream media (idnes and lidovky), owned by a Czech politician and millionaire. According to this HNA, partisan news media have nowadays a stronger position in Czech news media landscape than in Slovakia.

4.2 Are the Linking Patterns Different in Traditional and Partisan Media? Who Are the Main Actors of the Networks?

Different linking patterns can be found between partisan news media and mainstream news media. The clusters of the hyperlinks from and to the mainstream media correlated with the ownership (publisher) of these media both in Slovak (Fig. 3) and Czech Republic (Fig. 4). There seems to be only one exception (hyperlinks between two media sme.sk and hnonline.sk of different ownership),

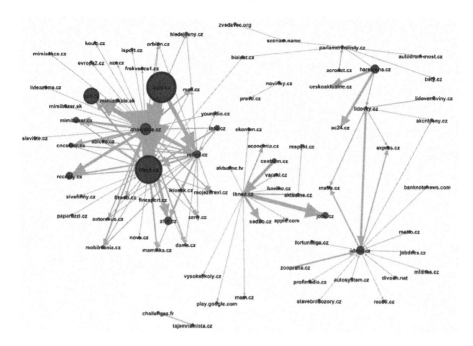

Fig. 2. Network of directed hyperlinks from and to the most popular Czech traditional and partisan news media

but a deeper analysis of Ahrefs results revealed that this number of hyperlinks was created due to redirections of one subdomain in one media to another. This was caused probably because of the transfer of ownership between these two subdomains. The other links were usually automatic interlinks between these portals. This result of individual clusters suggests independence and separateness of the researched mainstream media and/or unwillingness to link to the "competition".

Analyzed partisan media (both Slovak and Czech) create a denser and more interlinked network that confirms their interdependence in the network. Six co-cited (co-linked) Slovak alternative media nodes can be spotted in contrast with traditional Slovak media, where the network consists of usually separated clusters. The graph (Fig. 5) shows also partisan portals with highest importance in the network: dolezite.sk, infovojna.sk and hlavnespravy.sk. The first two link massively both to traditional and partisan media that is a very different linking pattern from traditional media that (with one above mentioned exception) rarely link to partisan media. Slovak partisan media position themselves as parts of the online news media landscape and the hyperlinks to the established media might at first glance look like an indication of their credibility. Nevertheless, this is valid only for Slovak domains, Czech alternative news media do not usually link to the mainstream media. In Czech partisan online news media landscape, the

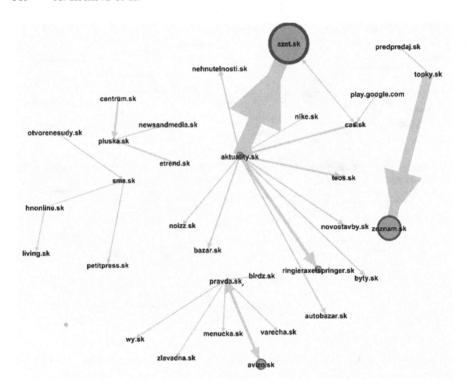

Fig. 3. Network of directed hyperlinks from and to the most popular Slovak traditional media

actors with highest importance (parlamentnilisty.cz and slobodnyvysilac.cz) are apparent similarly as in Slovak partisan online news media landscape (Fig. 6).

Besides links to news media (and their web hosting companies), most of the links from partisan news media point to the portals about boundary topics as bitcoin, green architecture and esotericism. Just the most popular alternative medium (hlavnespravy.sk) is attracting a wider audience with the links from music portal and portal for infographics/memes. The links from and to traditional media are more business oriented or targeted to a mainstream lifestyle. One of the traditional media (sme.sk) links regularly to their source that is a database of published court decisions.

If studying hyperlink relations within the sum of the most popular traditional news media in two countries (in Czech and Slovak republic), there is almost no significant hyperlink connection between Czech and Slovak traditional media portals, although some of them have the same publisher (owner). The explanation lies in a slight language barrier and different local news coverage of these media. These media are not interlinked by millions of hyperlinks and this fact suggests that they do not heavily transmit the content from one country to another. Figure 7 shows that clusters of traditional media of two

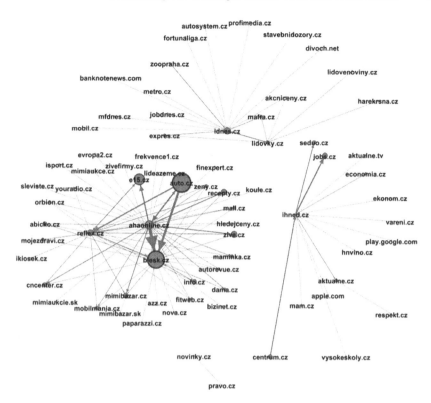

Fig. 4. Network of directed hyperlinks from and to the most popular Czech traditional media

countries (represented by .sk and .cz domains) remain separated except of one traditional medium (Pravda.sk) that shares the same publisher with one alternative medium.

On the other side, partisan news media of two independent states are much more likely to be interlinked also internationally than traditional media. This fact can be explained by transmitting the content from one media to another that is a well-known strategy of the partisan news media proved by the hyperlinks (Fig. 8).

5   Conclusions and Research Limitations

The hyperlink network analysis of mainstream and alternative online news media in two countries showed that hyperlinks are reliable indicators of an affinity of these media. Hyperlinks no longer serve just as citations between the portals and their presence only is not an indicator of the quality of the media, especially in the online news landscape. Different linking patterns and link building strategies of mainstream and alternative news media were demonstrated in this

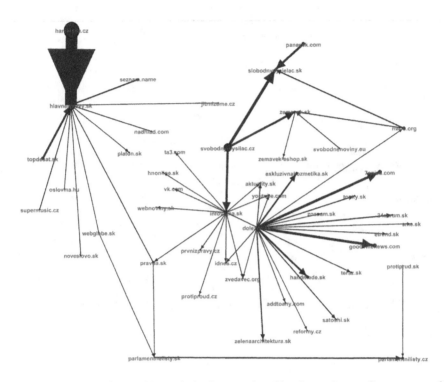

Fig. 5. Network of directed hyperlinks from and to Slovak partisan online news media

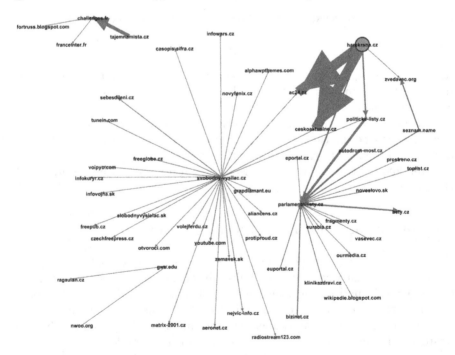

Fig. 6. Network of directed hyperlinks from and to Czech partisan online news media

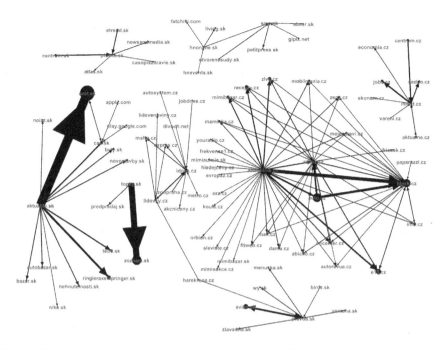

Fig. 7. Network of directed hyperlinks from and to Slovak and Czech mainstream online news media

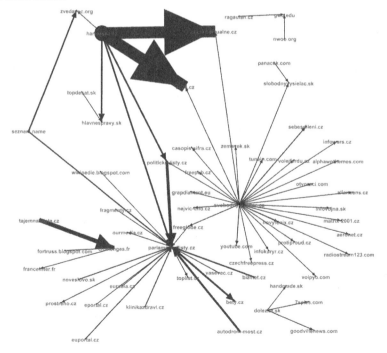

Fig. 8. Network of directed hyperlinks from and to Slovak and Czech partisan online news media

article. Mainstream online news media create hyperlink clusters within their own publisher. More context would be provided to the reader, if these news media of different publishers were interlinked. However, it is not the case of selected traditional media, and the advantage of this fact is that the ownership of these news media can be reliably inferred from the hyperlink structure. Another important conclusion is that mainstream online news media that do not share the same publisher with partisan media, never link heavily to the mainstream online news media of another state, even if they share the same publisher.

Partisan news media are much more interlinked between the clusters within one state, but also within two independent states. This fact suggests either their interdependence or partnership in the "partisan fight" against mainstream media or the same ownership as is the case of mainstream media. Nevertheless, if alternative and mainstream news media have the same owner, this fact is also visible in the interlinkage of these media. The position of partisan news media in the news media landscape differs in Czech and Slovak Republic. While Slovak partisan media link also often to the mainstream news media (it is not common vice versa with dofollow attribute), Czech partisan news media have a stronger position in the news media landscape and are not interlinked with different mainstream news media.

Two limitations of this research were identified. The first limitation is the focus on the online news media landscape of two states: Czech and Slovak Republic. However, this limitation is necessary, as the definition and scope of alternative and mainstream media differs in different countries and countries with a similar situation needed to be selected. Repeating this research with the same research methods, but in different countries is therefore suggested.

The second limitation is the restriction of the research to the links that are crawlable by robots. Some traditional media are under paywall and their links remain not crawled. Nonetheless, the hyperlinks to the affiliated media surely are crawlable and crawled by robots as hyperlinks with dofollow attribute serve also for the search engine optimization of these portals. Part of the sources, especially in alternative news media remain listed as pure text and part of sources are not stated at all. This proves that the hyperlink analysis does not serve as citation analysis of the sources, but rather as an analysis of cooperation between the selected actors.

As future work, we are interested in qualitative analysis and testing the identified features or linking patterns (e.g. links, dofollow links, etc.) as predictors of the news media credibility. Also, it is worth researching the spreading patterns of the information across the different media and different languages (e.g., English and Russian as the most relevant ones for the Central Europen media landscape).

Acknowledgments. This work was supported by the Slovak Research and Development Agency under the contract No. APVV SK-IL-RD-18-0004 (MISDEED) and APVV-17-0267 (REBELION).

Appendix

Overview and scope of Slovak partisan online news media analyzed in this article:

- www.dolezite.sk - News hub
- www.hlavnespravy.sk - Politics, army, interesting facts from technologies and popculture. Conservative targeting
- www.infovojna.sk - Politics, army
- www.parlamentnelisty.sk - Politics
- www.protiprud.sk - Politics, health, spiritual, conspiracies
- www.slobodnyvysielac.sk - Politics, conspiracies, culture
- www.zemavek.sk - Health, spiritual, "science", society

Overview and scope of Slovak mainstream online news media analyzed in this article:

- www.dennikn.sk - Politics, science, culture, sport, environmental issues
- www.hnonline.sk - Finance, politics
- www.pluska.sk - Showbusiness, politics, sport. Tabloid
- www.pravda.sk - Politics, finance, sport, culture, auto-moto, health and cooking
- www.sme.sk - Politics, economy, sport, culture, technologies, health, environmental issues
- www.topky.sk - Politics, showbusiness, leisure. Tabloid
- www.webnoviny.sk - Politics, finance, sport
- www.aktuality.sk - Politics
- www.cas.sk - Showbusiness, politics, sport. Tabloid

Overview and scope of Czech partisan online news media analyzed in this article:

- www.ac24.cz - Politics, army, "science"
- www.aeronet.cz - Politics, technologies
- www.ceskoaktualne.cz - Politics, health, culture, finance, army, conspiracies and factoids
- www.cz.sputniknews.com - Politics
- www.parlamentnilisty.cz - Politics
- www.pravyprostor.cz - Politics, society
- www.protiproud.cz - Politics, health, spiritual, conspiracies
- www.svobodnenoviny.eu - Anti-migrant, conspiracies, mysteries, health
- www.svobodny-vysilac.cz - Politics, spiritual, conspiracies, culture
- www.zvedavec.org - Politics, army, society, anti-LGBTI and migrants

Overview and scope of Czech mainstream online news media analyzed in this article:

- www.ahaonline.cz - Leisure, showbusiness. Tabloid
- www.blesk.cz - Leisure, showbusiness, politics. Tabloid
- www.ct24.ceskatelevize.cz - Politics, finance, culture, science, sport
- www.idnes.cz - Politics, finance, culture. Tabloid

- www.ihned.cz - Politics, finance, culture, technologies
- www.lidovky.cz - Politics, finance, culture, leisure
- www.pravo.cz - Politics, leisure, finance
- www.reflex.cz - Politics, interesting facts, culture
- www.respekt.cz - Politics, culture, society

References

1. Allcott, H., Gentzkow, M.: Social media and fake news in the 2016 election. J. Econ. Perspect. **31**(2), 211–36 (2017)
2. Björneborn, L., Ingwersen, P.: Toward a basic framework for webometrics. J. Am. Soc. Inf. Sci. Technol. **55**(14), 1216–1227 (2004)
3. Bowler, L., He, D., Hong, W.Y.: Who is referring teens to health information on the web? Hyperlinks between blogs and health web sites for teens. In: Proceedings of the 2011 iConference, pp. 238–243 (2011)
4. Chomsky, N., Herman, E.S.: Manufacturing Consent. The Political Economy of Mass Media. Pantheon, New York (1988)
5. Gelfert, A.: Fake news: a definition. Informal Log. **38**(1), 84–117 (2018)
6. Hrckova, A., et al.: Unravelling the basic concepts and intents of misbehavior in post-truth society. Bibliotecas. An. de Investigacion **15**(3), 421–428 (2019)
7. Jackson, M.H.: Assessing the structure of communication on the world wide web. J. Comput.-Mediated Commun. **3**(1), JCMC311 (1997)
8. Media nechcú, aby sme premýšľai. https://bit.ly/32m66Qs. Accessed 1 Sept 2019
9. Nemeth, M.: Traditional and alternative online news media in Hungary (e-mail communication), 20 July 2019
10. Park, H.W.: Hyperlink network analysis: a new method for the study of social structure on the web. Connections **25**(1), 49–61 (2003)
11. Park, H.W., Thelwall, M.: Link analysis: hyperlink patterns and social structure on politicians' web sites in South Korea. Qual. Quant. **42**(5), 687–697 (2008). https://doi.org/10.1007/s11135-007-9109-z
12. Richards, W.D., Barnett, G.A.: Progress in Communication Science. Ablex, Norwood (1993)
13. Severyn, A.: Traditional and alternative online news media in Poland (e-mail communication), 20 July 2019
14. Szabo, G., Bene, M.: Mainstream or an alternate universe? Locating and analysing the radical right media products in the Hungarian media network. Intersections. East Eur. J. Soc. Polit. **1**(1), 122–146 (2015)
15. Weigl: Celá európska integrácia prebieha pod zástavou oslabenia národných štátov a ich odumretia. https://bit.ly/2IR8aIj. Accessed 1 Sept 2019
16. Zeifman, I., Breslaw, D.: A Closer Look at the Most Active Good Bots. https://www.imperva.com/blog/most-active-good-bots/. Accessed 4 Sept 2019

Relations Between the Concepts
of Disinformation and the Fogg Behavior Model

Enrique Muriel-Torrado and Danielle Borges Pereira(✉)

Federal University of Santa Catarina, Florianópolis, Brazil
enrique.muriel@ufsc.br, danielle.borges.pereira@gmail.com

Abstract. This article aims to relate the concepts of false news and disinformation with the Fogg Behavior Model (FBM) in order to observe their existing connections. For this, two steps were carried out: a) Identify five concepts of fake news or disinformation of organs or institutions that work on information, education, science and/or society issues; and b) Relate the concepts of fake news or disinformation found with Fogg Behavior Model. The research has a qualitative approach, having characteristics of exploratory and documentary research. It is concluded that there were links between definitions of disinformation and fake news by UNESCO (United Nations Educational, Scientific and Cultural Organization), European Commission, Dictionary Oxford, World Economic Forum and Reuters with FBM with diverse levels of similarity. Through this analysis it is possible to learn more about how false information is formulated and how it can influence human behavior.

Keywords: Disinformation · Fake news · Fogg Behavior Model · FBM

1 Introduction

Many words are difficult to pinpoint their meaning, even when we use it daily and understand what they try to convey. Information is one such a word, in which multiple definitions apply to, in different situations and fields of study.

When it comes to conceptualizing, defining or giving meaning to the word information, the authors Capurro and Hjorland [20] affirm the existence of several concepts, from the most general to the most specific and when they are related to particular areas and disciplines. The importance of defining this word intends to solve existing problems in different fields, areas and disciplines; an example is the concept of information in the field of information sciences which, according to Losee, must have a general and precise definition [1].

Losee [1], in his study, defines information in a general way as: "information is produced by all processes and it is the values of the characteristics in the processes, output that are information". In treating information as a factor of oral and written communication, it has always been present in the course of the development and evolution of human beings, from the act of informing particular news or subject to the conveyance

R. Mugnaini (Ed.): DIONE 2020, LNICST 319, pp. 147–163, 2020.
https://doi.org/10.1007/978-3-030-50072-6_12

of knowledge and teachings. Developed, formulated and used by society according to their customs, needs and cultures [21].

Nowadays, with the arrival of political and intellectual combat, where population becomes its own threat, individuals think and manifest themselves in a more emotional than reasonable way. Nativism is gaining strength in numbers and society is on the right path for individualism. Politics and other correlated matters are becoming binary, instead of representing a dispute of ideas and relevant criticism. Science, studies and research are treat with doubt, unrest, suspicion and even contempt [2].

Those aspects end up favoring fake news that, as Alcott and Gentzkow [3] put it, are like articles on fake news published with the intent of informing untruth and misleading readers. Most of those articles are political biased, as shown in the United States presidential election of 2016. After this event, fake news began to invade our lives, along with post-truths (wherein emotions and personal beliefs value more than scientific facts), turning information into an object of manipulation, and transforming a lie into a truth in the eyes of those who have the propensity to trust [2].

Therefore, to comprehend the relationship between human's behavior with the internet and its offered tools, Fogg developed a psychological model called the Fogg Behavior Model (FBM), which "shows that three elements must converge at the same moment for a behavior to occur: Motivation, Ability, and a Prompt" [4].

Thus, this paper's objective is to identify if it exists a relation between fake news and disinformation's conceptions with the Fogg Behavior Model, in order to observe its existent connections. The research was done in two parts: a) identify five notions of fake news or disinformation in institutions or agencies that work on matters of information, education, science and/or society; and b) Relate the conceptions of fake news or disinformation found with the Fogg Behavior Model.

Studies made to understand what information and matters linked to it entails, such as disinformation and fake news, makes us improve our knowledge about the relevancy of researches done about the theme and recognize its importance for understanding and developing our society.

2 Information and Disinformation

Since the creation of the Gutenberg printing press in 1450, information is constantly undergoing development and improvement, from which knowledge becomes the social transformation of the population. Informational content is produced and disseminated on a large scale, which allows greater accessibility. Meanwhile, libraries have stopped being mere repositories and are used as study centers [5].

It was from this period, with the growing publication of news that began skepticism and questioning by intellectual authority, a process that led the scientific fields to produce their contents focused on scientificity [5]. Today we can see that the population constantly questions the authenticity and forgets to verify the information. In many cases people tend to believe news that brings up issues with which individuals choose to believe rather than scientificity.

Burke [5] considers that this informational explosion, aggravated by the emergence of the internet and social media (environments of easy publication and dissemination of news), brought us to the "knowledge crisis". That is, a world with a lot of information and disinformation, where the population no longer knows what to trust and end up choosing to believe in what is convenient for them. After this period, at the end of the twentieth century, we experienced with greater precision the transformation in our culture resulting from the technological paradigms involved in Information Technology (IT) and later in Information and Communication Technology (ICT).

In the Industrial Revolution, energy had a huge impact on society; people could have light in their homes, cook and do other jobs more easily. With the arrival of IT, it was as if there was a new revolution: the internet and its popularization changed and are still changing the lives of many people and the way they relate to others around them. Hence, Castells names this period the Technological Revolution, where Internet has a similar importance to what energy had in its emergence [6].

Castells relates the Industrial Revolution to the Technological Revolution based on changes and patterns of development in populations' economy, society and culture. In the first Industrial Revolution, science had no support, but information was widely used. In the second Industrial Revolution, science aided in the innovations of this period. Nowadays, in the Technological Revolution, the information and knowledge produced apply in society, creating new information and knowledge, forming a cycle of innovation and continuous use. However, this last revolution usually happens in a centralized way and in limited locations [6].

Another revolution presented by Castells is the Information Technology Revolution. It emerged amidst a new social structure, with cultural diversity and remodeling of the social behavior of communication forms, where it brings the informationalism that speaks about information networks, information flows and our dependence on them [6].

The growth of information has led several authors to start their research in order to understand this term. In the field of Information Science, Buckland [7] identifies information in three main uses of the word. Information-as-process deals with the act of informing, information-as-knowledge, seen in information-as-process, which is knowledge gained from communication. Lastly, information-as-thing, that designates objects, such as data and documents, from which it can provide knowledge.

Capurro, however, differentiates Information into three predominant epistemological paradigms in Information Science, which are the physical (physical object of a signal and the transmission of a message, must have informative value), cognitive (related to the search for knowledge due to a gap or need) and social (treats information as something linked to the user) [8].

D'Ancona addresses the revolutions that involve IT and/or ICT as Digital Revolution, which has brought many benefits to humanity, such as improved information democratization and access to tools that help us in everyday life, making it increasingly impossible to imagine a world without the use of technology, its software, applications and programs. Digital technology is one of the most innovative inventions in the world and the way it fits into people's lives quickly, practically and naturally is imperceptible. This has made the digital environment, especially the World Wide Web, a mirror of humanity [2].

Media modifications eventually aided the growth of a treacherous industry, beginning with the rising reach of high-speed broadband internet. This in turn, transformed it into a cheap communication and publication tool, fast and accessible as never before seen, bringing a behavioral and cultural impact on society as a whole [2].

These new forms of connections made people use social media to disseminate much more information than normally, consequently increasing the sharing of news without checking its veracity. Considering this in a larger scale, it is possible to say the sharing of fake or modified news has become a State issue, bringing with it risks for the population in a social, cultural and cognitive aspect [2].

Fake news and disinformation have been around for centuries, but were not as globalized as nowadays, with the use of the internet and sharing on social media. Today, fake news comes from various types of websites. Some of those are specifically designed to bring false information, some with the intent to make fun of it, while others aim to make people believe the news is true [3].

To understand how disinformation has become a threat in digital environments through social media, we will start with research and studies related to human behavior and see what leads individuals to believe, share and disseminate false or adulterated news.

Thus, the behavior must be clearly understood and its purpose known, "[…] whether is it being investigated a behavior (observable acts) or a behavioral consequence". Therefore, we are researching behavioral consequences, as well as human development, their personal, social and cultural relationships and how these factors may affect their future contribution to the growth of disinformation [9].

3   Fogg Behavior Model (FBM)

To make effective codifications that lead to changes in human behavior, especially in persuasive technology, it is necessary to have knowledge about human psychology and what guides this behavior. Therefore, FBM, as previously mentioned, is a psychological model that emerged from the need to understand the changes that occur in human behavior, especially when it comes to its communication channels and current technologies. With this model, it is possible to think in a systematic way about the behavioral changes in human beings [10].

This model is in constant development and improvement, as behavioral changes shifts from new social, cultural and technological interactions. FBM is based on persuasive technology that deals with "[…] learning to automate behavior change." FBM is divided into three main factors: motivation, ability and triggers, each of which has its own subcomponents. The use of this model characterizes the reasons that lead a person to perform such action, and for a specific human behavior to occur it is necessary to have the three factors present, initially having sufficient motivation and ability, as well as having an effective trigger that influence in this process [10].

To better understand how FBM occurs, Fig. 1 exhibits the two-axis model, presenting the relation between Motivation (vertical axis) and Ability (horizontal axis) in order to perform a target behavior, i.e. a desired behavior or behavior to be analyzed. The vertical axis corresponds to the motivation, being the upper part of the axis considered high motivation and the lower part low motivation of an individual to perform the pre-established behavior. In the horizontal axis it deals with the element Ability, where the left side corresponds to low Ability and the right side to high Ability of an individual to execute the target behavior [10].

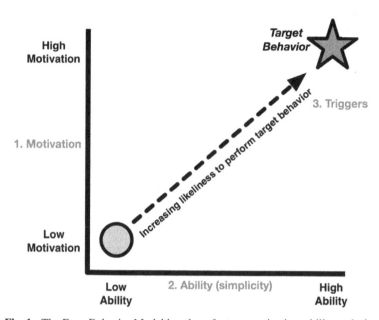

Fig. 1. The Fogg Behavior Model has three factors: motivation, ability and triggers [10].

In addition to the axis presented by this model, there is the factor that Fogg names triggers, an item that is close to the star (represents the behavior of the target). The presence of this factor is essential for the behavior to occur. The trigger and star can be directed anywhere, differently from the fixed axes of motivation and ability. In the case of motivation and ability they are usually necessary for behavior to occur, but as depicted in Fig. 1 with the diagonal arrow, it emphasizes the relationship between motivation, ability and target behavior (star). Individuals with high skills and motivation are more likely to perform the desired behavior [10].

Motivation and ability are factors that do not necessarily have to be high or low at the same frequency, but do need to be related. People with high motivation and low ability can perform desired behavior, just as individuals with low motivation but high enough ability to accomplish their goal. What cannot happen is an individual with no motivation or ability. Often, the increase in ability is more important than motivation, since the former refers to the facilitation of target behavior, and these levels are likely to be of manipulation [10].

The third factor raised by Fogg is the trigger which can be considered the most important one. It is through an appropriate trigger that the effectiveness of the desired behavior is defined. When a trigger is "[…] successful, it has three characteristics: First, we notice the trigger. Second, we associate the trigger with target behavior. Third, the trigger happens when we are motivated and able to perform the behavior" [10].

In order for the design of a target behavior to be able to develop such activity efficiently and effectively, that is, a well-formulated and designed behavior that leads to its result with the user fulfilling the behavior it is necessary to analyze the target user and the target behavior. This analysis is conducted to determine the best subsets of the motivation, ability and trigger factors to be used [10].

Motivation is mainly responsible for its elements that lead a user to have enough of it to perform certain behavior. Motivation has three pairs of elements that are presented in Table 1, the use of these items aims to raise the user's position to high motivation [10].

Table 1. Elements of motivation

Motivation (elements of motivation)	
Pleasure/Pain	Usually immediate factor, little thought or anticipation, are powerful motivators and a primitive response, being able to adapt to various situations. To increase motivation, pleasure and pain may be incorporated. It is not considered an ideal approach, but it must be recognized as a motivation option
Hope/Fear	Anticipation of a result. Hope is the anticipation of something good happening and fear is the anticipation of something bad. People are motivated by hope (joining a dating site) and fear (updating antivirus software). Hope isprobably FBM's most ethical and empowering motivator
Social Acceptance/Rejection	Controlling part of our social behavior, people are motivated to do things that give them social acceptance and even more motivated to avoid being socially rejected. Our track record shows that being banned from a community is a form of punishment. With social media, methods to motivate people through social acceptance or rejection has flourished. People are significantly motivated by the desire to be socially accepted

Source: Based in Fogg [10].

The ability to deal with the ease of performing the desired behavior is divided into six parts Fogg calls the "elements of simplicity" presented in Table 2 which, when used together, yield the best results. This item is not about teaching the user how to perform a task, but about simplifying that activity so that they are more likely to do it [10].

The last factor in FBM is the trigger that may have different nomenclatures, but is intended to make an individual perform a desired behavior immediately. The triggers have three forms of which are presented in Table 3, but not all people are stimulated to perform a behavior by the same triggers and often make us act on impulse momentarily [10].

The purpose of this model developed by Fogg is to understand the human behavior when performing some activities, especially those that are pre-established. Their results can explain the factors that lead individuals to perform certain functions and how digital

Table 2. Elements of simplicity.

Ability (elements of simplicity)	
Time	If target behavior takes time and we have no time available, the behavior may not occur
Money	For people with limited financial resources, targeted behavior that costs money may not happen
Physical Effort	Behaviors that require physical effort may not be simple
Brain Cycles	For the most part, we overestimate how much ordinary people want to think. If performing a target behavior leads us to think too much, it may cause the individual not to achieve the desired
Social Deviance	If a behavior requires going against the norm, violating the rules of society, that behavior is no longer simple
Non-Routine	Activities and behaviors that need to get out of routine, people may not find it simple and end up not performing the desired behavior

Source: Based in Fogg [10].

Table 3. Three types of triggers.

Triggers (three types of triggers)	
Spark as trigger	Used when a person has no motivation to perform a behavior and so, a trigger that has to do with a motivational element is designed. Sparks can leverage any of the motivational elements and can be accomplished in many ways, as long as the trigger is recognized and presented to users at the moment they can act
Facilitator as trigger	Suitable for users who are highly motivated but not skilled, the facilitator triggers the behavior and at the same time makes it easier. An effective facilitator tells users that target behavior is easy to perform, that it does not require a feature that he or she does not have at the moment
Signal as trigger	This type of trigger works best when people have the motivation and ability to perform the behavior. The signal does not seek to motivate people or simplify the task, it just serves as a reminder

Source: Based in Fogg [10].

developers can use tools to systematically stimulate, persuade or manipulate stipulated behaviors, which involve human feelings [10].

Then, by presenting the concepts of disinformation in organs and institutions that deal with information, we will approach these definitions relating to the factors and their subcomponents presented by Fogg, in order to understand if the concepts of disinformation have characteristics that can lead humans to spread fake news through motivation, ability and/or triggers.

4   Methodological Procedures

The data collection procedures make it a documentary research, aiming to identify five conceptions of fake news or disinformation by agencies or institutions working on information, education, science and/or society issues. This research has a qualitative approach, being an exploratory research aiming to relate the concepts of false news or misinformation developed and provided by UNESCO (United Nations Educational, Scientific and Cultural Organization), European Commission, Oxford Dictionary, The World Economy Forum and Reuters with the Fogg Behavior Model.

The study plans to contribute to the Information Science field's theoretical development in which it discusses the behavioral relationships of humans with information and disinformation, seeking to understand this phenomenon and therefore, identify manners to prevent the growth in disinformation at environments that deal with information and communication media.

5   Analyze

The vast amount of information that is produced today is disorganized and this has caused some professions working with communication and information to lose its quality, truth and credibility, for example journalism [11].

For UNESCO much of this information is being published and disseminated in order to spread falsehoods, especially on social networks. With this the organization classified and defined these types of news as: disinformation, misinformation and misleading information. Still stating the contradiction of calling this news Fake News, since if a news is false, it cannot be considered news [11].

The definitions of false information for UNESCO [11] are divided into three items:

- Disinformation: Information that is false and deliberately created to harm a person, a social group, an organization or a country.
- Misinformation: Information that is false but not intended to cause harm.
- Misleading Information: Reality-based information used to harm a person, social group, organization, or country.

Another more simplified definition of "Fake News" for UNESCO states that "[…] is intentionally lying content that is launched on fake websites or social networks to harm a person or a group. UNESCO prefers to call them disinformation campaigns." [12].

The European Commission has defined disinformation as "[…] verifiable, false or misleading information created, presented and disseminated for economic gain or intentionally misleading the public." Disinformation may undermine the credibility and trust of some institutions, In addition to the power to undermine our democracy with consequences for the entire population, causing public damage to politics, health, safety and the environment, it also makes it difficult for individuals to gain access to truthful information and impedes freedom of opinion and expression [13].

In the Oxford Dictionary [14] it has the definition of the term's fake news, disinformation and post-truth that basically deal with fake news and its spread and the power that individuals' beliefs can bring about whether they believe it or not. Your definitions are:

- Fake News: false reports of events, written and read on websites. Many of us seem unable to distinguish fake news from the verified sort. Fake news creates significant public confusion about current events.
- Misinformation: the act of giving wrong information about something; the wrong information that is given. A campaign of misinform. Ation attitudes based on misinformation.
- Post-truth: relating to circumstances in which people respond more to feelings and beliefs than to facts. In this era of post-truth politics, it's easy to cherry-pick data and reach any conclusion you like.

The World Economy Forum [15], a non-profit organization defined Fake News: "can be as slippery to define as it is to pin down. Stories may be factually inaccurate and deliberately published to underscore a certain viewpoint or drive lots of visitors to a website, or they could be partially true but exaggerated or not fully fact-checked before publication." For Reuters [16] British news agency, one of the largest international agencies in the world, defines Fake News from the viewpoint of its readers as: "[…] fake news is only partly about narrowly defined fabricated news, and very more about broader discontent with the information landscape - including media and politicians, as well as companies and platforms."

It is important to know how to identify and deal with fake news, but especially in virtual environments the information provided is poor and misleading and with some intent behind it, and it is this disinformation most read and found by people [16].

The following tables present the relations linking concepts of misinformation and false news as defined by selected institutions and agencies with the motivational factors defined by Fogg, which include: Motivation (Pleasure/Pain, Hope/Fear and Social Acceptance/Rejection), Ability (Time, Money, Physical Effort, Brain Cycles, Social Deviance and Non-Routine behavior) and Triggers (Spark, Facilitator and Signal).

Table 4 presents the UNESCO definitions of Disinformation, Misinformation, and Malicious Information, the FMB elements that relate to the definitions, and a brief explanation of this relation.

The definition of Disinformation related to the element of Motivation: Pleasure/Pain that refers to false information trying to affect certain groups through their feelings and beliefs, and the element Trigger: Spark, complementing the element of Motivation when the individual has a motivation to believe in determined subject, sometimes all it takes is a spark, a fake news, that meets this person's expectations in order for them to believe in it without checking its veracity.

The definition of Misinformation relating to the element of Motivation: Social Acceptance/Rejection, because it deals with false information but not created to harm, usually is disseminated in order to individuals acquire social acceptance from a particular group, or not telling the factual news for fear of social rejection. The definition of Malicious Information is related to the element Ability: Social Deviance, as it is true information,

Table 4. Relation: definitions UNESCO and elements FBM.

Institution	Term used	Definitions	Elements FBM	Relation: Definitions and Elements FBM
UNESCO	Disinformation	Information that is false and deliberately created to harm a person, a social group, an organization or a country	Motivation: Pleasure/Pain	The definition of misinformation as an information intended to harm a person or a group is related to the elements of Motivation pleasure/pain, as it tries to affect certain groups through their feelings, creating triggers for greater commotion
			Trigger: Spark	The Trigger as Spark is the false information created to harm, when the trigger combined with the pleasure/pain element is projected, this factor can affect certain people or groups
	Misinformation	Information that is false but not intended to cause harm	Motivation: Social Acceptance/Rejection	This definition can be related to the element Motivation: Social Acceptance/Rejection, as it deals with the necessity of informing or believing in something to be socially accepted or failing to inform or believing in something for fear of social rejection
	Misleading information	Reality-based information used to harm a person, social group, organization, or country	Ability: Social Deviance	This definition deals with true information that is stolen and spread, most of the times illegally, with the purpose of harming a person or group. This definition is related to element Ability: Social Deviance that deals with a behavior that goes against the standards and rules in order to cause damage

Source: Research Data, 2019.

but generally retrieved through illicit means and disseminated with the intention of creating harm, causing the individual's behavior to fit into Social Deviance, requiring going against social norms.

Table 5 exhibits the definition of Disinformation from European Commission, the FMB elements that relate to the definitions and a brief explanation on these relations.

Table 5. Relation: definitions European Commission and elements FBM.

Institution	Term used	Definitions	Elements FBM	Relation: Definitions and Elements FBM
European Commission	Disinformation	Verifiable, false or misleading information created, presented and disseminated for economic gain or intentionally misleading the publicmay undermine the credibility and trust of some institutions. In addition to the power to undermine our democracy with consequences for the entire population, causing public damage to politics, health, safety and the environment, it also makes it difficult for individuals to gain access to truthful information and impedes freedom of opinion and expression	Motivation: Social Acceptance/Rejection	These types of information impedes the freedom of information and expression that relates to element Motivation: Social Acceptance/Rejection, being that people are ashamed (social rejection) for informing "news" that could be fake or the necessity of informing any news, veracious or not, with the intent of being socially accepted
			Ability: Brain Cycles	This definition can be associated with the Ability: Brain Cycles meaning that it makes difficult to access to true information, as ordinary people usually don't have the habit of investigating certain information and news
			Ability: Time and Money	Generally, the false information are of easy access and understanding, this factor can be related to the element Ability: Time and Money, since it often brings brief and free information to the people
			Ability: Social Deviance	By spreading false information that attack the democracy affecting the whole population, the definition relates to the element Ability: Social Deviance, that requires going against the norms and rules.
			Trigger: Facilitador	The definition of disinformation refers to the easily accessed and comprehended (easy language) false news that relates to the element Trigger: Facilitator that aims to facilitate people's access to information, its understanding and even dissemination

Source: Research Data, 2019.

European Commission's disinformation definition is related to the Motiviation: Social Acceptance/Rejection element by defining that false information precludes liberty of opinion and expression, and that individuals may feel coerced into believing and disseminating certain information and emitting others so that they are socially accepted and avoid social rejection. Ability elements are the ones that best fit this definition and they are: Brain Cycles (by making it difficult to access true information), Time (for bringing false information quickly and briefly, saving time) and Money (for bringing false information free) and Social Deviance (by trying to achieve democracy with false information).

Table 6 presents the definitions of Fake News, Misinformation and Post-Truth developed and published by Oxford Dictionary, FMB's elements associated to the definitions and a brief explanation on these associations.

The definition of Fake News has relation to the Brain Cycles Ability element, because it may require a lot of thinking in order to find veracity in an information and Social Deviance, when talking about false event reports, being a fabricated information requires going against society's norms and rules.

Misinformation is linked to the Social Deviance Ability due to the act of providing false information being considered wrong and requires going against society's rules

Table 6. Relation: definitions Oxford Dictionary and elements FBM.

Institution	Term used	Definitions	Elements FBM	Relation: Definitions and Elements FBM
Oxford Dictionary	Fake News	False reports of events, written and read on websites. Many of us seem unable to distinguish fake news from the verified sort. Fake news creates significant public confusion about current events	Ability: Brain Cycles	The definition addresses public confusion, since it's common for the population to be unable to distinguish the true from the false and identify the truth of news, this factor may be related to the element Ability: Brain Cycles, in order to identify the truth of Information it might require a lot of thinking
			Ability: Social Deviance	Dealing with fake reporting, this concept is about the Social Deviance Ability, when the dissemination, or creation, of false information requires going against the norms and rules
	Misinformation	The act of giving wrong information about something; the wrong information that is given. A campaign of misinform. Ationattitudes based on misinformation	Ability: Social Deviance	Misinformation's definition deals with informing or reporting something wrong and it relates to the Social Deviance Ability element, as it requires going against society's norms and rules
	Post-truth	Relating to circumstances in which people respond more to feelings and beliefs than to facts. In this era of post-truth politics, it's easy to cherry-pick data and reach any conclusion you like	Motivation: Pleasure/Pain and Hope/Fear	This definition relates to the Hope/Fear and Pleasure/Pain from the Ability element, as it involves people's feelings and beliefs, giving them more importance than to the facts
			Trigger: Spark	Relating to Trigger: Spark, the definition acknowledges that when people believe in certain things it gets easier to believe in the information and when a Spark of grand commotion arises regarding a certain subject, it makes it more relevant and people answer to their feelings and beliefs more than to the facts

Source: Research Data, 2019.

and norms. The definition of Post-Truth relates to the Motivation: Hope/Fear and Pleasure/Pain elements, for its definition involves individuals' feelings, beliefs and ideologies, and may cause them to believe false information for the predisposition in believing in what is being informed and giving hope that such information is true.

Post-Truth definition relates to the Spark and Facilitator Trigger element, for when people present the easiness to believe in a certain information it gets simpler to make it happen once a spark occurs by giving the subject big commotion, making it is easier to the individual believe in what they want to believe.

Table 7 presents the definition of Fake News by the World Economic Forum, the FMB elements that relate to the definitions, and a brief explanation on that.

Fake News definition is linked to the Pleasure/Pain Motivation element, since it's a false information originated from an individual's point of view, perhaps addressing subjects that cause a predisposition to the user's approval as truthful, even when they might be fake and Social Acceptance/Rejection because it deals with a false information that affects the point of view of the individual that might be being influenced by social acceptance or rejection.

Ability element: Time, on directing users to websites with false information, sparing their time. And lastly, the Trigger element: Facilitator, by directing the user to a website with just one click, facilitating their contact with false information and a lot of times, interfering their access to true information, and Signal, that could only serve as a reminder of a certain subject that the individual already has the predisposition to believe in.

Table 8 presents the links that can be found between the definition of Fake News conceived by Reuters and the FMB elements, also a short explanation on their relations.

Fake News' definition fits only two elements, the first one is Motivation: Pleasure/Pain on saying that those news happen because of the discontentment of a scenario and certain false information can bring people to act instinctively through their beliefs and ideology. The second element is the Trigger: Spark that along with the Motivation: Pleasure/Pain, can make this individual believe even more in certain information.

From this analysis it was possible to understand the relationship of concepts with FBM, we can identify some characteristics that are used by designers when trying to stimulate a target behavior with the objectives of fake news. Basically the formulation of false information uses some mechanisms of ability, Motivation and triggers, especially the last two factors, which can make it easy for the individual to share false information with just one click, motivating the individual to perform such task with emotional, sentimental and ideological issues and raising triggers by showing discontent or anger about a given subject previously analyzed and pre-arranged by the user.

By relating the amount of times the elements appear, the Motivation presented eight occurrences, four of those were Pleasure/Pain, one Hope/Fear and three were Social Acceptance/Rejection. The Ability element also appeared eight times, two in Time, one in Money, two in Brain Cycles, three in Social Deviances and none in Physical Effort and Non-Routine Behavior. The element Trigger occurred six times, three of those as Spark, two as Facilitator and one as Signal.

Table 7. Relation: definitions World Economic Forum and elements FBM.

Institution	Term used	Definitions	Elements FBM	Relation: Definitions and Elements FBM
World Economic Forum	Fake News	Can be as slippery to define as it is to pin down. Stories may be factually inaccurate and deliberately published to underscore a certain viewpoint or drive lots of visitors to a website, or they could be partially true but exaggerated or not fully fact-checked before publication	*Motivation: Pleasure/Pain and Social Acceptance/Rejection*	This definition informs that the false news is published with the intent of exalting a certain point of view and it correlates to the Pleasure/Pain Motivation element that may address a determined subject that the users might possess predisposition to believe in. Also correlating to the Social Acceptance/Rejection Motivation element, since the users' points of view are often related to their social behavior and their worries on being accepted or rejected in a certain social group
			Ability: Time	Fake news directs the users to specific websites, this factor connects to the Time Ability element because if the user's desired target behavior saves them time, this behavior is easier to happen, for example reading news with only a click
			Trigger: Facilitator	This factor relates to the Trigger element: Facilitator as it facilitates user access to fake news with just one click, knowing that fake news directs users to specific websites
			Trigger: Signal	This definition can be related to the element Trigger: Signal, as it can serve as a reminder of some subject from certain point of view, and along with some element of motivation and ability, it becomes more effective

Source: Research Data, 2019.

Table 8. Relation: Definitions Reuters and Elements FBM.

Institution	Term used	Definitions	Elements FBM	Relation: Definitions and Elements FBM
Reuters	Fake News	Fake news is only partly about narrowly defined fabricated news, and very more about broader discontent with the information landscape - including media and politicians, as well as companies and platforms	Motivation: Pleasure/Pain	This definition is related to the Motivation element: Pleasure/Pain and Hope/Fear when it comes to fake news, presenting partially from a discontent of a specific scenario, generally involved with users' beliefs and ideologies
			Trigger: Spark	The definition of Fake News acknowledges news that bring discontentment from a specific scenario, also relate to the Trigger element: Spark, being that the user that sees afake news already has involvement with the subject and with just a spark they become more likely to believe in such information

Source: Research Data, 2019.

Considering that the elements that related the most to the definitions developed by the agencies and institutions presented were Motivation and Ability, when dealing with sub-elements, the ones that had the highest occurrence within each element were Motivation: Pleasure/Pain, Skill: Social Deviance and Trigger: Spark as trigger.

Knowing that for Fogg [10] a target behavior to be successfully performed, they must involve all Motivation, Ability and Trigger elements, so false information must have at least one subdivision of each element.

Given the growing proportion of disinformation and fake news, UNESCO, the European Commission, IFLA (International Federation of Library Associations and Institutions) and FackCheck are some of the institutions that make materials available to prevent and identify false information. UNESCO talks about the damage caused by false information and how it can harm our personal lives and society as a whole in relation to politics, health, violence, rights and social exclusion. Therefore, to combat disinformation, you should only share content on the Internet when you are sure it is truthful and have critical thinking, questioning the information before disseminating it [11]. The European Commission brings as a way to combat disinformation on social media a set of activities carried out to all involved, from social media platforms, institutions, individuals, among others [13]. IFLA's Fake News prevention can be considered one of the most shared and well-known by internet users, it presents ways and questions we must do to identify if a news is false or true. How to raise some questions about the news: is it a joke? Affect beliefs or ideologies? Do you receive any support that may influence the report? When was it published (many old stories spread like new ones)? What is the source and author? In addition, you need to pay attention to attention-grabbing news and if you still have any questions about the news, consult an expert (librarian) or verification websites [17].

These ways of identifying fake news presented by IFLA were based on the FactCheck [18] article published in 2016. In addition to the items we must be aware of presented by IFLA, FactCheck still has a need to verify the support on which the information is and if it is some kind of joke, being news intentionally false such as satire.

6 Conclusion

During the process of interaction between humans, we have a need to influence or change each other's thoughts and behaviors, from parents wanting their children to do what they want, to salespeople who want to sell products to customers. This phenomenon is linked with social psychology, called social influence where one person tries to induce another to a desired behavior, but this is just one of many ways to change human behavior. There is also behavior change in conjunction with attitude change that involves new personal beliefs and preferences [19].

Both of these behaviors' changes can fit FBM and in consequence to the creation, publication and sharing disinformation. As seen in Tables 4, 5, 6, 7 and 8, the definitions presented are linked to the FBM factors exhibited in the Tables 1, 2 and 3, some on a large scale such as Oxford Dictionary (six sub-elements) and European Commission (six sub-elements), followed by Word Economic Forum (five sub-elements) and UNESCO (four sub-elements) and others on a small scale, like Reuters (two elements).

However these definitions hint that fake news develops a positive or negative motivation for the reader to share it, a skill to disseminate it with just a click sharing it in social media, and also have several triggers that could be understood as scandalous headlines or as biased towards beliefs and ideologies that drive the individual to share it.

The relation presented in this paper aims to show that the concepts related to false information comply with all the factors developed by Fogg in his behavior model. Through this analysis it is possible to learn more about how false information is formulated and how it can influence human behavior. Hopefully with the conclusion of this research, there will be some development of future studies on human behavior and how individuals deal with false information, especially in digital environments, in order to seek effective mechanisms to solve the dissemination of disinformation.

References

1. Losee, R.M.: A discipline independent definition of information. J. Am. Soc. Inf. Sci. **48**(3), 254–269 (1998). https://ils.unc.edu/~losee/book5.pdf
2. D'Ancona, M.: Pós-verdade: a nova guerra contra os fatos em tempos de fake news. Faro Editorial, São Paulo (2018)
3. Alcott, H., Gentzkow, M.: Social media and fake news in the 2016 election. J. Econ. Perspect. **31**(2), 211–236 (2017). https://www.nber.org/papers/w23089
4. Fogg, B.J. https://www.behaviormodel.org/
5. Burke, P.: Uma história social do conhecimento: de Gutenberg a Diderot. Zahar Ed., Rio de Janeiro (2003)
6. Castells, M.: A sociedade em rede, 7th edn. Paz e Terra, São Paulo (2003)
7. Buckland, M.: Information as thing. J. Am. Soc. Inf. Sci. **42**(5), 351–360 (1991)
8. Capurro, R.: Epistemologia e Ciência da Informação. V Encontro Nacional De Pesquisa Em Ciência da Infomração. UFMG, Belo Horizonte (2003). http://www.capurro.de/enancib_p.htm
9. Moutinho, K., Roazzi, A.: As teorias da ação racional e da ação planejada: relações entre intenções e comportamentos. AvaliaçãoPsicológica **9**(2), 279–287 (2010). http://pepsic.bvsalud.org/scielo.php?script=sci_arttext&pid=S1677-04712010000200012
10. Fogg, B.J.: A behavior model for persuasive design. In: Proceedings of the 4th International Conference on Persuasive Technology – Persuasive 2009 (2009). https://doi.org/10.1145/1541948.1541999
11. UNESCO. http://www.unesco.org/new/pt/brasilia/communication-and-information/freedom-of-expression/media-development/disinformation/
12. UNESCO. http://www.unesco.org/new/fileadmin/MULTIMEDIA/FIELD/Brasilia/pdf/brz_ci_what_are_faknews_por_2019_01.pdf
13. European Comission. https://ec.europa.eu/digital-single-market/en/tackling-online-disinformation
14. Oxford Dictionary. https://www.oxfordlearnersdictionaries.com/
15. Charlton, E.: Fake news: what it is, and how to spot it. World Economic Forum (2019). https://www.weforum.org/agenda/2019/03/fake-news-what-it-is-and-how-to-spot-it/
16. Nielsen, R.K., Graves, L.: "News you don't believe": audience perspectives on fake news, pp. 1–8. Reuters (2017). https://reutersinstitute.politics.ox.ac.uk/sites/default/files/2017-10/Nielsen&Graves_factsheet_1710v3_FINAL_download.pdf
17. IFLA. https://www.ifla.org/publications/node/11174
18. Kiely, E., Robertson, L.: How to spot fake news. FactCheck (2016). https://www.factcheck.org/2016/11/how-to-spot-fake-news/
19. Rodrigues, A.: Psicologia social para principiantes: estudo da interação humana, 13th edn. EditoraVozes, Rio de Janeiro (2011)
20. Capurro, R., Hjorland, B.: O conceito de informação. Perspectivas em Ciência da Informação **12**(1), 148–207 (2007). https://doi.org/10.1590/S1413-99362007000100012
21. Freire, G.H.: Ciência da Informação temática, histórias e fundamentos. Perspectivas em Ciência da Informação **11**(1), 6–19 (2006). http://www.scielo.br/pdf/pci/v11n1/v11n1a02

Evaluation of Science in Social Networking Environment

A Strategy for Co-authorship Recommendation: Analysis Using Scientific Data Repositories

Felipe Affonso$^{(\boxtimes)}$, Thiago Magela Rodrigues Dias,
and Monique de Oliveira Santiago

Centro Federal de Educação Tecnológica de Minas Gerais, Belo Horizonte, Brazil
felipe-affonso@hotmail.com, thiagomagela@gmail.com,
moniqueosantiago@gmail.com

Abstract. In a co-authorship network papers written together represent the edges, and the authors represent the nodes. By using the concepts of social network analysis, it is possible to better understand the relationship between these nodes. The following question arises: "How does the evolution of the network occur over time?". To answer this question, it is necessary to understand how two nodes interact with one another, that is, what factors are essential for a new connection to be created. The purpose of this paper is to predict connections in co-authorship networks formed by doctors with resumes registered in the Lattes Platform in the area of Information Sciences. To this end, the following steps are performed: initially the data is extracted, later the co-authorship networks are created, then the attributes to be used are defined and calculated, finally the prediction is performed. Currently, the Lattes Platform has 6.1 million resumes from researchers and represents one of the most relevant and recognized scientific repositories worldwide. Through this study, it is possible to understand which attributes of the nodes make them closer to each other, and therefore have a greater chance of creating a connection between them in the future. This work is extremely relevant because it uses a data set that has been little used in previous studies. Through the results it will be possible to establish the evolution of the network of scientific collaborations of researchers at national level, thus helping the development agencies in the selection of future outstanding researchers.

Keywords: Co-authorship networks · Scientific data repositories · Lattes Platform

1 Introduction

In the late 1990s, several researchers devoted attention to network studies. Work has been done on biology, the internet, routers, among others [4, 19, 22]. From this

Supported by CAPES.

R. Mugnaini (Ed.): DIONE 2020, LNICST 319, pp. 167–178, 2020.
https://doi.org/10.1007/978-3-030-50072-6_13

moment on, social networks became the focus of research. Work has also been carried out on various types of networks to understand their properties and characteristics [20]. Based on this it was possible to represent them mathematically, which further boosted the progress of the works that aimed to analyze the characterized networks. Metrics, theories and indices were adopted to measure the behavior of the networks. Work has also been done to differentiate social networks from non-social networks [22].

From the analysis of networks, it is possible to explain several phenomena. Social network analysis allows us to understand the relationship between nodes. Studying these links between nodes for a while raises the question, "How does the evolution of the network occurs over time?", understanding the evolution of the network as a whole is a complex task [3].

With these concepts in mind, the link prediction problem was proposed [12]. Initially, methods were used to calculate the similarity between two network nodes. The more similar the nodes, the more likely they are to be linked together.

Therefore, several other methods have been proposed to better solve the prediction problem of links [1,14,27]. Probabilistic, linear algebra-based, and binary classification methods were proposed, thus, several algorithms can be used for its resolution. In this paper, we will treat the prediction of links as a classification problem, thus, algorithms in recommendation systems area are used to achieve the proposed objectives.

Applying such concepts to a more specific domain, we can turn our attention to networks belonging to the scientific community. When publishing a paper with another scientist, a connection is formed by the collaboration made. In these networks the authors represent the nodes, and the scientific collaborations represent the edges [16]. Such networks are called co-authorship networks, and will our main object of study.

In this context, the Lattes Platform, maintained by the CNPQ[1], has been a source of data from various works aimed at analyzing scientific collaboration networks, mainly because it encompasses data from much of the national scientific production. Lattes Platform currently has 6.1 million researcher curricula and represents one of the world's most relevant and recognized scientific data sources [11]. The data in the curricula registered in the Lattes Platform has attributes such as: name, academic background, professional experience, projects, scientific publications, among others. The sheer volume of data in curricula can provide valuable and up to now unknown information [7].

Understanding the evolution of the network requires understanding how two nodes interact with each other. The network is formed by the relationship between the nodes, so we seek a way to predict which researchers will produce a joint article in the future. Such behavior is present basically in all social networks through the "suggested friends". Thus, it is possible to use the same techniques for the scientific collaboration networks studied in this work.

Given the arise of recommendation systems, which represent a specific approach to machine learning concepts. By employing this technique it is possible

---

[1] National Council for Scientific and Technological Development.

to understand which attributes of the nodes make them closer to each other, and thus have a greater chance of creating a relationship in the future.

Therefore, the prediction of links in co-authoring networks formed by the data of doctors with curricula registered in the Lattes Platform in the area of Information Science will be performed. With this, it will be possible to understand the behavior of the network and monitor its evolution over time. Through this study, it will also be possible to identify the researchers who can collaborate in a future instant of time. In a second moment, starting from the proposed analysis, it also becomes possible to identify the most influential researchers in the co-authorship network.

The text is organized as follows: Sect. 2 presents the works related to this as well as the definition of some concepts that are important for the execution of the work. The Sect. 3 presents the methods used, explaining all the techniques and decisions taken to complete the work. The results obtained from this methodology will be presented in Sect. 4. Finally, a conclusion and some future work are presented in Sect. 5.

2   Literature Review

In a seminal work, the link prediction problem, as we know is defined [12]. This study is still considered the starting point for this field. The theme is introduced focusing on social networks and their dynamism. Over time, new edges are added to the networks, which represents the emergence of new interactions in the social structure. The authors define the problem of link prediction as: given a social network at a time t, the goal is to accurately predict the edges that will be added to the network during the interval t and a time future t '. Link prediction, in this context, allows one to discover individuals who are already working together, but their interaction has not yet been directly observed [10].

With this same goal in mind, a study aiming to discover which source of information could indicate relationships between users [2]. Throughout this work, several steps are taken to understand the connection of one user with another. In this paper, the author refers to the problem as "relationship prediction", and uses a ranking of similar people to predict the missing edges. At the end of the study, a portion of the students were given a list of people most similar to them, and often recognized such individuals. The author points out that the great challenge of such analyze is to have only a small data set, which represents a tiny portion of the actual data.

However, in order to predict a missing link, concepts related to the topological characteristics of the network must be better understood. To this end, a work that focuses on analyzing the main differences between social and non-social networks is conducted [22]. It highlighted that the relationship between the degrees of the adjacent nodes of the networks are positively correlated in social networks, but negatively in other types of networks. Secondly, social networks show a high level of clustering. In conclusion, social networks are divided into communities, while non-social networks are not. In this context, we can understand the degrees of a network as the minimum distance, in terms of numbers of areas in the network, between all pairs of nodes in the network, through which a connection exists [19].

Even after several studies in the area of social network analysis, understanding the entire evolution of a network is a complex task, but understanding the association between two specific nodes is much simpler [3]. Therefore, some questions may be asked: How does the pattern of associations change over time? What are the factors that guide these associations? How is the association between two nodes affected by other nodes? To answer the questions, the author uses the standard problem formulation [12] and conducts a survey of existing approaches focusing mainly on social network graphs.

Turning the attention to the networks of scientific collaboration, object of study of this work, [21] presents one of the first works on this topic. Three specific networks are studied, one in biomedical research, one in physics, and lastly, mathematics. The author presents several characteristics of co-authorship networks, and performs several analyzes to understand the behavior of nodes in this network. The importance of such networks is highlighted, and how they have meticulous, well-documented information and even temporal events in the social and professional relationship of scientists.

Using the Lattes Platform as a data source, an approach for extracting researchers' curricula and building a scientific collaboration network is described [7]. The relationship between employees is accomplished through the presence of one or more works together. Through the built framework, networks that have common terms, participated in the same congress or even in the same area are presented. In [6], the authors present the method in detail. Some tests are performed and the properties present in them is analyzed.

An approach aiming to find most influential researchers in a collaborative network is presented [25]. For this, a link predictor based on local metrics of the network structure is used. The collaborative individual influence is obtained by taking into account the influence of a particular researcher on the prediction of network links as a whole. The data from 47,555 Lattes Platform researchers curricula are used, which were obtained using ScriptLattes [17]. As a result, the measures of collaborative influence present a significant inverse correlation when compared to the most well-known centrality measures. This fact demonstrates the effectiveness of the proposed metrics. Another important factor is that the described methodology can be calculated independently for each vertex, without the need for a global calculation, thus reducing the computational cost [24].

3   Methodology

In order to achieve the proposed objectives, some steps are necessary. This section will highlight the methods used to predict future connections in a specific area. Therefore, the large area of Social and Applied Sciences was chosen, and later the sub-area Information Science. This data set has 1,094 PhD researchers curricula. Initially, the framework used for data extraction will be presented.

Secondly, the scientific collaboration networks created, and lastly, the attributes selected for the prediction will be characterized.

To begin the development of the work, it was necessary to perform the extraction of the data to be used. For this, the *LattesDataExplorer* [8], a framework for data extraction and processing was used. As can be seen in Fig. 1, initially the data is collected through CNPq and stored in a local repository where data selection is performed. Using the identifier of each curriculum, the date of the last update is compared with the repository in CNPq. If the dates are different, the extractor replaces the curriculum that was stored locally with the most current version [8]. Afterwards the data is processed and stored in XML (Extensible Markup Language) format, so that it is possible to generate metrics and calculate some statistics.

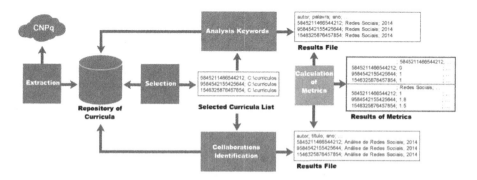

Fig. 1. Framework used for data extraction [8].

With the data extracted and organized, it is necessary to create the networks. The co-authorship of an article can be understood as the documentation of a collaboration between two or more authors, and these collaborations form a "network of scientific collaboration" [21]. A method for identifying scientific collaborations in large databases using low computational cost was used to generate the networks used in this work [5].

After collaboration networks are created, it is necessary to identify which attributes will be used for prediction. Therefore, a basic set of features from other works that addressed this theme were selected [2,3,9,12,15,24].

The simplest way to perform edge prediction is through the common neighbors metric [12], which can be understood as the number of common nodes that two specific nodes have. Using this attribute in scientific collaboration networks, it is pointed out that individuals who have never worked together but have a common collaborator are much more likely to collaborate in the future [23]. The Common Neighbors (CN) attribute is demonstrated in Eq. 1, where x and y represent vertices of the graph.

$$CN(x,y) = |\Gamma(x) \cap \Gamma(y)| \qquad (1)$$

Another metric that can be obtained using the structural characteristics of the network itself is called Jaccard Coefficient (JC), and measures the probability that both x and y have a v neighbor, randomly chosen that x or y own. Unlike the Common Neighbors attribute, the Jaccard Coefficient normalizes the number of common neighbors [3], as follows:

$$JC(x, y) = \frac{|\Gamma(x) \cap \Gamma(y)|}{|\Gamma(x) \cup \Gamma(y)|} \tag{2}$$

In order to establish similarity between two pages, Adamic/Adar metric is proposed [2]. In order to use it in link prediction algorithms, it was customized and it is presented in Eq. 3 [13]. This formulation gives the rarer characteristics a greater weight [26]. We can understand it as the number of properties shared by nodes, divided by the log of the frequency of the characteristics.

$$Adamic/Adar(x, y) = \sum_{w \in \Gamma(u) \cap \Gamma(v)} \frac{1}{\log |\Gamma(w)|} \tag{3}$$

Following the same reasoning, the Resource Allocation (RA) metric assigns weight to the two-node relationship favoring relationships between those with few relationships [9], and can be found in Eq. 4.

$$RA = \sum_{w \in \Gamma(u) \cap \Gamma(v)} \frac{1}{|\Gamma(w)|} \tag{4}$$

Considering only the size of the node neighborhoods, the Preferential Attachment (PA) metric has been proposed, and is presented in Eq. 5. In short, it establishes that the probability of a new relationship with other vertices is based on the degree of the node in question [3]. This metric does not require neighborhood-related information for each node, so it has a lower computational cost [15].

$$PA = |\Gamma(u)||\Gamma(v)| \tag{5}$$

The fact that friends of friends can create a connection suggests that the distance between nodes in a network can influence the formation of new connections [3]. In this way, the Shortest Path (SP) metric can also be used in order to predict links. We can understand it as the minimum path between two nodes [9].

Domain-related attributes can also be used during the prediction process. In this case, it is necessary to evaluate the data set used and the techniques necessary to convert them to the correct formats for input to the algorithm. By using the Lattes Platform, various information present in the curriculum of researchers, such as: orientations made, participation in newsstands, congresses in which some publication was held, institutions where the researcher studied, among others. In the present work, as the Information Science sub-area is already being used, the data were used: city, state and institution.

The use of categorical data involves a coding process to enable its use in the prediction process. Therefore, each of the aforementioned information was coded by two processes: LabelEncoding and OneHotEncoder, using libraries already developed for this purpose. From the transformations performed, each categorical data became a sparse matrix.

Finally, the number of joint collaborations that two nodes had over that time was also considered as an attribute. This way it is possible to identify researchers who have been working together longer, and possibly have a greater influence in the next few moments.

3.1   Proposed Method

After defining the attributes that will be used, some steps are necessary. Firstly it is necessary to define the periods for training and testing, so 3 different networks were created. For network 1, the publications made between 1960 and 2000, which will be called the initial period, were defined. The second network was created for the period from 2001 to 2010. Finally, the period from 2011 to 2018 was established for the third and last network. Such periods include the date of the first work registered on the platform until the last year prior to the presentation of this work. Figure 2 presents the 3 networks, it is possible to observe that over time the collaborations between the scientists increased. Through this representation it is also possible to understand the purpose of the work in question. Given the first network, predict what the collaborations will be in the next instant of time.

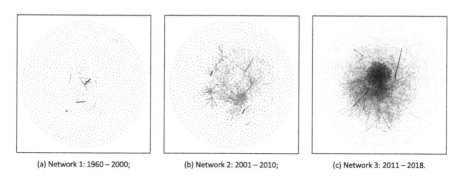

(a) Network 1: 1960 – 2000;   (b) Network 2: 2001 – 2010;   (c) Network 3: 2011 – 2018.

Fig. 2. Evolution of the scientific collaboration network of the information science subarea

Table 1 presents the main characteristics of the created networks. The amount of edges increases considerably over the years. The attributes mentioned above were identified from the use of such networks, most of them use topological attributes to calculate the metric. It is also important to note that the number of researchers did not change over time, the same group of nodes was selected for the entire period.

Table 1. Key networks properties

Network	Period	Number of nodes	Number of edges	Centrality
Network 1	1960–2000	1084	191	0.3524
Network 2	2001–2010	1084	1537	2.8358
Network 3	2011–2018	1084	3831	7.0683

The data set containing the researchers, the links between them and the selected attributes was then used as input to a machine learning algorithm. Each row in the data set is composed of the following items: First Researcher Identification, Second Researcher Identification, Common Neighbors (CN), Jaccard Coefficient (JC), Adamic/Adar (AA), Resource Allocation (RA), Preferential Attachment (PA), Shortest Path (SP), weight, City, State, Institution, and finally the presence or absence of an edge. It is important to note that the indices correspond to the calculations previously presented for the two nodes of the line. The edge is obtained using data from the later period. That is, given this set of attributes, will a new edge be generated? This information will be sent to the prediction algorithm.

At this stage of the work, the problem of class imbalance comes up. The number of possible links in a graph is quadratically related to the number of nodes, however, the number of actual links represents only a small fraction of this number [3]. This problem interferes with the results due to two reasons: (i) with fewer examples of a given class, it is more difficult to infer reliable patterns; (ii) trained models are skewed towards the predominant class [18]. Several authors [1,3,25] propose techniques and methods for solving this challenge. A traditional technique for overcoming class imbalance is called over-sampling. It consists of reducing the number of samples of the determinant class randomly, thus equating the number of components for both cases. This technique was used in the work presented here. Initially the data set had a ratio of 152 missing edges for each edge present. After over-sampling, the number of edges present and absent is the same. With balanced data, the prediction algorithm was executed.

4   Results

Throughout the process described in the previous section, the dataset has undergone some changes. The 1,084 researchers can have a total of 587,528 edges. Of these, only 3,831 represented positive edges in Network 3, so by balancing the samples, a random set of other 3,831 missing edges was chosen. Thus, the data set used for input to the prediction algorithm is made up of 7,662 records. Thus, 5,746 (randomly chosen) lines were selected for the training part, representing 25% of the total set, and another 1,916 were selected for the test part.

Table 2. Metrics generated from predictions

Algorithm	Precision	Recall	F1	AUC
Support Vector Machine	0.78	0.77	0.77	0.86
Logistic Regression	0.78	0.78	0.78	0.87
K-Nearest Neighbors	0.74	0.73	0.73	0.80
Naive Bayes	0.77	0.63	0.58	0.63
Random Forest	0.77	0.77	0.77	0.85

Several algorithms can be used to solve classification problems, among them, some were selected to perform the work, they are: Support Vector Machine, Logistic Regression, Nearest K-Neighbors, Naive Bays and Random Forests. Each of these techniques has a different peculiarity and, consequently, different results. Therefore, their results will be presented in Table 2, using the metrics precision, recall, F1 and area under the curve (AUC). Usually, in link prediction algorithms, the area under the curve is used by most authors, so we will use it as basis.

Each of the metrics used to validate the results has its own characteristics. Accuracy aims to answer the following question: Of all positive predicted values, how many are actually correct, high accuracy is related to a fewer false positives. Already considering all the positive values, the recall aims to know how many of these were actually predicted. The F1 metric takes precision and recall into account, thus making a weighted average of these two metrics. Finally, the area under the curve (AUC) is used to present the performance of a classification model throughout the learning process.

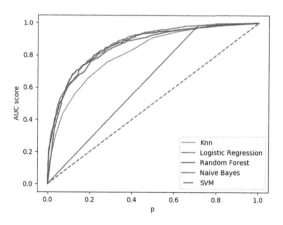

Fig. 3. Area under the curve (AUC)

Analyzing Table 2, it is possible to notice that the chosen algorithms obtained good results. Looking at the area under the curve, we realized that everyone got a result above a mere chance. This situation is better explained in Fig. 3, where the blue dotted line represents a 50% chance of hit, which means equal odds for the prediction to be correct or incorrect, and the orange line represents the values of the predictions made. Thus, it is clear that the algorithm was able to use the presented data set to make correct predictions about future connections.

Among the algorithms used, which presented the best performance, taking into account all metrics, was the Logistic Regression, followed by Support Vector Machine, Random Forests, , K-Nearest Neighbors, and, lastly, Naive Bayes. However, there is a slight difference between the results obtained, making it clear that, for the problem in question, we cannot yet establish which technique should be used as a standard.

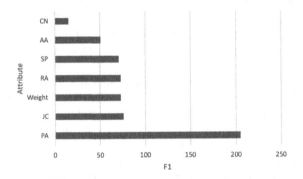

Fig. 4. Feature importance

By analyzing the learning process taking into account the topological attributes used (city, state and institution were not taken in account for this evaluation), it is possible to identify the order of influence of each one of them in the final result. We can observe from Fig. 4 that the order of importance of the attributes for the prediction performed here is: Preferential Attachment, Weight of Collaborations, Shortest Path, Jaccard Coefficient, Resource Allocation, Adamic/Adar, and, finally, Common Neighbors. This fact presents a behavior different from most of the theoretical references studied here, where, most of the time, the most relevant attribute is the Common Neighbors. However in the studies performed here, the Preferential Attachment metric is responsible for a good part of the final result.

5   Conclusion

The results presented here show that it is possible to perform the prediction of links using information from the network itself. The proposed objective was then achieved, since by using these data it is possible to know, for example, if two researchers from the area mentioned above will collaborate in a future instant of time. The performance of the evaluation metrics was around 80% representing a good result, but higher values can be achieved by using more attributes, mainly domain-related ones. It is possible to use the methods presented here to support decision making when granting scholarships, determining research groups and promoting researchers. Although the presented methods can be easily applied to similar studies, one of the major limitations found is the inability to replicate the works found in the literature regarding link prediction, since data sets are not public.

As future work, we highlight the importance of increasing the data set, or even looking for other ways to solve the problem of class imbalance, thus increasing the number of samples present for training the algorithm. From this, the classifiers are expected to perform even better.

References

1. Acar, E., Dunlavy, D.M., Kolda, T.G.: Link prediction on evolving data using matrix and tensor factorizations. In: ICDMW 2009. IEEE International Conference on Data Mining Workshops, 2009, pp. 262–269. IEEE (2009)
2. Adamic, L.A., Adar, E.: Friends and neighbors on the web. Soc. Netw. **25**(3), 211–230 (2003)
3. Al Hasan, M., Zaki, M.J.: A survey of link prediction in social networks. In: Social Network Data Analytics, pp. 243–275. Springer (2011). https://doi.org/10.1007/978-1-4419-8462-3_9
4. Barabási, A.L., Albert, R.: Emergence of scaling in random networks. Science **286**(5439), 509–512 (1999)
5. Dias, T.M.R., Moita, G.F.: A method for the identification of collaboration in large scientific databases. Em Questão **21**(2), 140–161 (2015)
6. Dias, T.M.R., Moita, G.F., Dias, P.M., Moreira, T.H.J.: Identificação e caracterização de redes científicas de dados curriculares. iSys-Revista Brasileira de Sistemas de Informação **7**(3), 5–18 (2014)
7. Dias, T.M., Moita, G.F., Dias, P.M., Moreira, T., Santos, L.: Modelagem e caracterização de redes científicas: um estudo sobre a plataforma lattes. In: BRASNAM-II Brazilian Workshop on Social Network Analysis and Mining, pp. 10–20 (2013)
8. Dias, T.: Um estudo da produção científica brasileira a partir de dados da plataforma lattes. 181p. Programa de Pós-Graduação em Modelagem Matemática e Computacional, Centro Federal de Educação Tecnológica de Minas Gerais, Belo Horizonte (Doutorado) (2016)
9. Digiampietri, L., Maruyama, W.T., Santiago, C., da Silva Lima, J.J.: Um sistema de predição de relacionamentos em redes sociais. In: Brazilian Symposium on Information Systems, vol. 11 (2015)
10. Krebs, V.E.: Mapping networks of terrorist cells. Connections **24**(3), 43–52 (2002)
11. Lane, J.: Let's make science metrics more scientific. Nature **464**(7288), 488 (2010)
12. Liben-Nowell, D., Kleinberg, J.: The link-prediction problem for social networks. In: Conference on Information and Knowledge Management (CIKM 2003), pp. 556–559 (2003)

13. Liben-Nowell, D., Kleinberg, J.: The link-prediction problem for social networks. J. Am. Soc. Inf. Sci. Technol. **58**(7), 1019–1031 (2007)
14. Liu, Z., Zhang, Q.M., Lü, L., Zhou, T.: Link prediction in complex networks: a local Naïve Bayes model. EPL (Europhys. Lett.) **96**(4), 48007 (2011)
15. Lü, L., Zhou, T.: Link prediction in complex networks: a survey. Phys.: Stat. Mech. Appl. **390**(6), 1150–1170 (2011)
16. Maruyama, W.T., Digiampietri, L.A.: Co-authorship prediction in academic social network. In: Anais do V Workshop Brasileiro de Análise de Redes Sociais e Mineração, pp. 79–90. SBC (2019)
17. Mena-Chalco, J.P., Junior, R.M.C.: Scriptlattes: an open-source knowledge extraction system from the lattes platform. J. Braz. Comput. Soc. **15**(4), 31–39 (2009)
18. Menon, A.K., Elkan, C.: Link prediction via matrix factorization. In: Gunopulos, D., Hofmann, T., Malerba, D., Vazirgiannis, M. (eds.) ECML PKDD 2011. LNCS (LNAI), vol. 6912, pp. 437–452. Springer, Heidelberg (2011). https://doi.org/10.1007/978-3-642-23783-6_28
19. Newman, M.E.: The structure of scientific collaboration networks. Proc. Nat. Acad. Sci. **98**(2), 404–409 (2001)
20. Newman, M.E.: Mixing patterns in networks. Phys. Rev. E **67**(2), 026126 (2003)
21. Newman, M.E.: Coauthorship networks and patterns of scientific collaboration. Proc. Nat. Acad. Sci. **101**(suppl 1), 5200–5205 (2004)
22. Newman, M.E., Park, J.: Why social networks are different from other types of networks. Phys. Rev. E **68**(3), 036122 (2003)
23. Newman, M.: Networks: An introduction. Oxford University Press, Oxford (2010)
24. Perez Cervantes, E.: Análise de redes de colaboração científica: uma abordagem baseada em grafos relacionais com atributos. Ph.D. thesis, Universidade de São Paulo (2015)
25. Perez-Cervantes, E., Mena-Chalco, J.P., De Oliveira, M.C.F., Cesar, R.M.: Using link prediction to estimate the collaborative influence of researchers. In: 2013 IEEE 9th International Conference on eScience (eScience), pp. 293–300. IEEE (2013)
26. Potgieter, A., April, K.A., Cooke, R.J., Osunmakinde, I.O.: Temporality in link prediction: understanding social complexity. Emerg.: Complex. Organ. (E: CO) **11**(1), 69–83 (2009)
27. Zhou, T., Lü, L., Zhang, Y.C.: Predicting missing links via local information. Eur. Phys. J. B **71**(4), 623–630 (2009)

A Distributed Tool for Online Identification of Communities in Co-authorship Networks at a University

David Fernandes[1], Nuno David[1,2(✉)], and Maria João Cortinhal[1,3]

[1] University Institute of Lisbon – ISCTE-IUL, Lisbon, Portugal
nuno.david@iscte-iul.pt
[2] Dinamia-CET ISCTE-IUL, Lisbon, Portugal
[3] CMAF-CIO, Faculdade de Ciências da Universidade de Lisboa, Lisbon, Portugal

Abstract. Most universities have their public repositories of scientific publications available online. The data is made available raw or by department listing and does not provide the network of co-authorships that implicitly emerges from scientific collaborations among different departments. Sometimes, the network of co-authorships is computed within the institution, via standalone applications that have few or no functionalities to explore the structure of collaborations. The possibility of searching online and managing the network of scientific communities in the institution is a matter of management efficiency, both for the institution itself and other external collaborators. This paper explains a distributed architecture and a tool that uses data from an online institutional repository. The tool calculates and puts available online the co-authorship network that identifies research communities according to different algorithms. The tool reflects and identifies the emergent structure of communities, graphically analyses communities, exports, reports and follows up with the evolution of communities in time.

Keywords: Online institutional repositories · Community detection tools · Interdisciplinary collaboration · Co-authorship · Graph · Author · Publication · ABCD and MCL algorithms

1 Co-authorship Networks in ISCTE-IUL

Interdisciplinary collaboration fosters researchers to combine collective expertise and gain synergies. This may result in increased productivity, originality and higher-impact research. In academic institutions, most internal interdisciplinary, collaborative research emerges from ad hoc groups, based on personal relations. However, most institutions are structured into discipline-focused, encapsulated units, such as departments, schools and faculties. The segmentation of institutions into sub-structures of decentralized scientific areas, with higher or lower degrees of autonomy, is justified by organizational and financial efficiency. Without adequate instruments to foster collaboration, organizational segmentation is likely to limit the potential of collaboration among actors, either

© ICST Institute for Computer Sciences, Social Informatics and Telecommunications Engineering 2020
Published by Springer Nature Switzerland AG 2020. All Rights Reserved
R. Mugnaini (Ed.): DIONE 2020, LNICST 319, pp. 179–189, 2020.
https://doi.org/10.1007/978-3-030-50072-6_14

internal or external to the institution, making it more difficult to foster inter- and multi-disciplinary ideas and initiatives. Perceiving and analyzing collaborative research in academic institutions can be a way to develop academic policies for promoting further interdisciplinary research (Newman 2004).

Most universities have their public repositories of scientific publications available online. However, the network of co-authorships is usually computed within the institution, via standalone applications, with few or no online functionalities to explore and provide the structure of collaborations. In this work, we are interested in the development of tools that allow users and faculty boards to perceive the emerging social network structure of co-authorships and explore online the potential space of further collaboration in their institution. Automatically identifying communities enables detection of patterns of knowledge sharing within academic institutions, which otherwise would be imperceptible. In order to do so, we implemented a tailored framework that allows examining the evolution and the structure of the co-authorship network revealed by scientific publications of authors of ISCTE-IUL Lisbon University Institute (ISCTE-IUL)[1].

The framework identifies communities in the co-authorship network, according to different algorithms. It uses online data from the institutional repository database to construct the co-authorship graph, calculate the communities graphs, storing it into a documented oriented database and a making it available online to interested users.

Community detection algorithms is a graph clustering problem. Community detection is automatic, without human intervention, in the sense that the purpose of the algorithm is to mathematically find groups of nodes with higher likelihood of connecting each other than to nodes from other communities. However, given the availability of different algorithms in the literature, and the need to empirically analyze how and why users identify themselves with the generated communities, it is vital to provide online tools to facilitate the incorporation of new algorithms in the framework.

The purpose of this paper is to substantiate and describe the framework design and its ability to generate and provide online different networks based on different algorithms. Main requirements of design included: distribution, where users can access online the graph and analyse co-authorships from a web site; persistence, where the graph is analysed at a particular time, communities are detected and stored, making it possible to access it in a responsive way the evolution of the emergent networks through time; modularity, where new algorithms for community detection can be added and capture alternative views of communities.

The structure of the paper is as follows. In the next section we recall the concept of co-authorship networks. In Sect. 3 we describe the framework architecture and in Sect. 4 the implemented algorithms. In Sect. 5 we describe the results of the authorship community detection in ISCTE-IUL, according to two algorithms, and a give brief comparative analysis of results. Finally, in Sect. 6, we present the conclusions.

---

[1] ISCTE-IUL is a public university with approximately 10 000 students (https://www.iscte-iul.pt). It comprises four schools, 8 research centers and more than 500 professors and researchers.

2   Co-authorship Networks

Co-authorship networks are social networks that have been widely studied to determine the structure of scientific collaborations (Jackson 2008). Unlike citations, co-authorships involve a temporal and collegial relationship, and consequently places them more squarely in the realm of social network analysis (Liu et al. 2005).

A co-authorship network can be modelled by an undirected graph in which nodes and edges represent authors and co-authorship relationships, respectively. Besides the number of shared scientific publications, there are other factors that are important to shape collaboration patterns among authors. For instance, if one article has two authors and another article has ten authors, the authors in the first article should be considered more connected than those of the second article. To express this relationship magnitude we considered a weighed network as follows.

Let $G = (V, E, W)$ be the co-authorship graph G where V is the set of nodes (authors), E is the set of edges (co-author relationships between authors), and W is the set of weights w_{ij} associated with each edge connecting a pair of authors (v_i, v_j). The weight of each edge (v_i, v_j) is then given by:

$$w_{ij} = \sum_{k=1}^{K_{ij}} \frac{1}{(n_k - 1)} \tag{1}$$

where K_{ij} and n_k represent the number of scientific publications co-authored by at least v_i and v_j, and the total number of co-authors of such publications, respectively. According to (1), each co-authored publication adds to the co-authorship relationship the factor $\frac{1}{(n_k - 1)}$ (Newman 2001). From now on, these weights will be named as Newman attractiveness.

3   Framework Architecture

Data for constructing the network originates from Ciência-IUL (2018). Ciência-IUL is the institutional repository of scientific publications produced by the authors of ISCTE-IUL. The information present in Ciência-IUL must be transformed into a network of co-authorships where community identification algorithms can be applied. The overall framework is intended to provide users with online, easy access through a web browser, as illustrated in Fig. 1.

The framework architecture contains four modules, according to Fig. 2: co-authorship graph generation, community identification, database and website. The construction of the co-authorship graph is a time-consuming process. Thus, once the graph is generated it is persisted in the database. The modular nature of the solution allows future replacement of either module without compromising the use of the remaining modules. For example, a new community identification algorithm can be added, which uses the persisted graph in the database without any changes to the modules responsible for generating the graph.

Data are collected from Ciência-IUL through a REST[2] API provided by Ciência-IUL and transformed into a graph with authors as vertices and co-authorships as edges

---

[2] Representational State Transfer (REST).

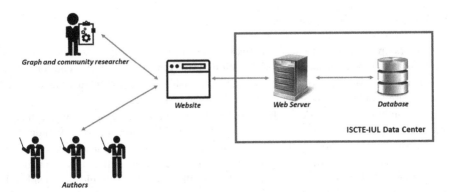

Fig. 1. Framework architecture.

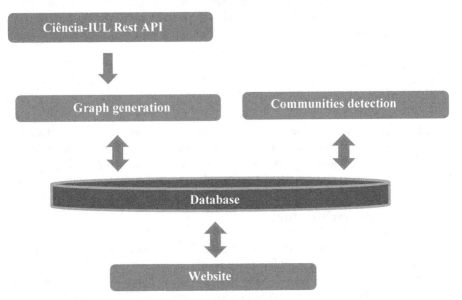

Fig. 2. Solution architecture.

between them. The REST API offers what are called endpoints to connect, gather information or perform some functionality in the Internet. Such web-services allow requesting systems to access and manipulate textual representations of web resources using a uniform, predefined set of stateless operations.

Like most publication repositories, Ciência-IUL API does not search for entities that have changed in a given period. This requires searching for all entities every time a co-authorship network is calculated. This limitation implies unbearable processing times, which would preclude providing online the networks in a responsive way. The solution was to construct a non-relational database (Fig. 3) with MongoDB (MongoDB 2017) that stores every graph. To consume the Ciência-IUL endpoints and create the co-authorship network we used the Node.js software. (Node.js 2017). The framework is thus able to consult and compare communities calculated over time.

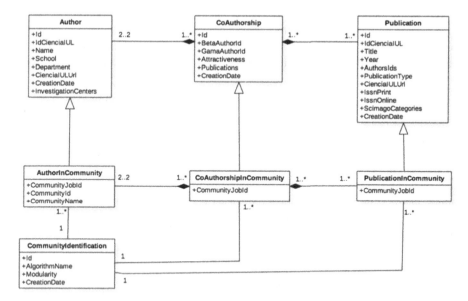

Fig. 3. UML data entities.

Once the graph is generated – a simple network with no identified communities – it is persisted in the database and may then be accessed through the graphical interface, in the website. Community identification algorithms can be applied and their results also stored, persisted in the database and accessed through the same website.

The Community Identification component is where community identification algorithms are executed. To the extent that it is a separate component, which will have its own timing and logic, one can easily add or remove algorithms to the solution. There are currently two algorithms in place, ABCD and MCL.

The graphical interface, through the website, allows to visualize online the co-authorship graph, as well as the communities identified in a commercial web browser. Its essential features are the visualization of the graph with the co-authorship network, the visualization of a graph with several identified communities (all vertices of the same community are the same color), the ability to query the communities or author details by accessing the vertex(es) that it represents and a set of descriptive metrics (such as author numbers, number of publications, number of co-authors, etc.). Exporting graphs in standard format to other general-purpose tools, such as GEXF - Graph Exchange XML Format (GEXF Working Group 2018), is also provided.

The database has the structure shown in Fig. 3. Each co-author has two authors and at least one publication, represented by the Publication entity. Whenever an instance of publication is created its Scimago scientific categories are added in the ScimagoCategories property. A publication may be in several different co-authorings. Each of these entities has a set of properties that characterize it in the context of the co-authoring network. Note that this database is not intended to be a copy of Ciência-IUL. In this context, in order to know more details about an author or a publication the repository Ciência-IUL may be consulted.

Whenever communities are identified at a given time the community graph is stored in the derived entities AuthorInCommunity, CoAutorshipCommunity and Publication-InCommunity. The context of community identification, as well as the algorithm used, are stored in the CommunityIdentification entity.

4   Unfolding Co-authorship Communities

The detection of communities in graphs is a problem with vast literature and it is normally known as a graph clustering problem (Liua et al. 2014). On a broad sense, a community is a group of nodes with higher likelihood of connecting each other than to nodes from other communities. How to determine that a node belongs to a community and not to another one is central in this problem. We used Newman's concept of attractiveness, in order to give us the likelihood of two nodes belonging to the same community. We implemented two communities detection algorithms, the Markov Cluster Algorithm (MCL) and Attractiveness-based community detection (ABCD). It should be stressed that other algorithms can be implemented, providing the online user with different ways to calculate communities.

4.1   MCL

The Markov Cluster Algorithm (MCL) follows the principle that a cluster of nodes has many edges inside it and few connections to other clusters (Dongen 2000). This means that if two nodes, u and v, are in the same cluster, the probability (the Newman force) that a path between u and v has external nodes to the cluster should be low. Therefore, a random walk between u and v has very little probability of leaving their cluster, in other words, their community. Random walks in a graph can be described by means of Markov chains in which the sequence of variables in the chain is represented by a sequence of probability transition matrixes.

Once the initial matrix is determined, the algorithm proceeds iteratively until the stopping criteria is verified, that is, until there is no difference between two consecutive transition matrices. In each iteration, a new matrix is calculated using two operators: Expansion and Inflation. The Expansion changes the transition probabilities so that they reflect the introduction of intermediate edges in the random walk from any vertex j to any vertex i. The Inflation operator, in turn, increases the difference between the highest and lowest transaction probabilities and, in this way, it reinforces the attractiveness of the stronger edges and reduces the attractiveness of the weakest edges.

Stijn van Dongen created MCL and provided a tool to use it (Dongen 2017). We used his implementation in our application of MCL. We used it with the operation Expansion with $p = 2$ and with the operation Inflation with $r = 2$.

4.2 ABCD

Attractiveness-based community detection, ABCD (Ruifang Liua 2014) is an algorithm used to detect communities in weighted graphs. It relies on two main concepts: density of a group of nodes and attractiveness between a group of nodes. It is an algorithm for agglomeration of nodes, where they are put in the same group until some condition is met and the agglomeration stops. It begins with as many groups as nodes. As the number of nodes in a group grows, the denser it becomes and the more difficult it is to merge with another. Two groups of nodes are merged together whenever their attractiveness is higher than their own density. The attractiveness is based on the weight of the edges, which is measured by the Newman attractiveness. A group density is the sum of the weights of its nodes, and the weight of a node is the average weight of its edges. In this way, there is a direct relation between authors and co-authorships.

5 Results

In this section, we summarize and analyse the scientific collaborations among ISCTE-IUL authors. Our purpose in this section is not fundamentally theoretical but to describe, as a proof of concept, the kind of information a user can access and explore through our distributed framework. We considered all the ISCTE authors that have co-authored at least a scientific publication with another ISCTE author from 1975 to March 2017, which comprises a 42 years interval. It is worth to remark that only per-reviewed scientific contributions were considered in this study.

5.1 The ISCTE-IUL Co-authorship Network

Co-authorship networks document scientific collaborations, where nodes are authors and a link represent the fact that two authors have written at least one scientific publication together. Thus, they are undirected graphs.

The ISCTE-IUL co-authorship network – see Fig. 4 – has 613 nodes and 1718 edges. To highlight multidisciplinary scientific collaborations the shade of each node depends on the school that each author belongs. Among the 613 nodes there are 92, 185, 195 and 141 nodes representing authors belonging to, respectively, the School of Technology and Architecture, the School of Sociology and Public Policy, the School of Social Sciences and the Business School.

Despite having 1718 edges, the ISCTE-IUL co-authorship represents 3766 scientific publications, since some pairs of authors have co-authored more than one scientific publication. However, the width of each edge in Fig. 4 does not represent the number of publications. It represents the Newman attractiveness force (see Eq. 1).

In Fig. 4 it may be observed that there is one school (Business School) in which, unlike other schools, authors share many co-authorships with other schools' authors but they do not form a strong block: they are a bridge that connects the remaining three schools. It could be theorized that authors of this school exert a knowledge sharing that tends to be more transversal to ISCTE-IUL.

By analyzing Newman attractiveness force, it was also possible to conclude that there is a huge difference among the top 5 co-authorships, 109 versus 20. Moreover, the

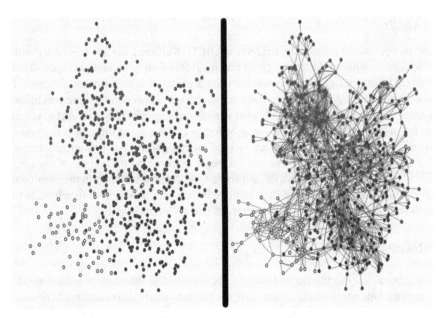

Fig. 4. ISCTE-IUL co-authorship network.

attractiveness of around 70% of co-authorships is equal to one, which reveals that the same pair of authors tend to share scientific publications only once. For this reason, the differences among the width of the edges in Fig. 4 are almost imperceptible.

5.2 Community Detection

As previously mentioned, two algorithms for community detection were implemented: MCL and ABCD. On what follows, communities with less than four authors were not considered and removed from this analysis. This was due to considerations of simplicity. However, this condition can be easily removed.

The MCL method identified 26 co-authorship communities in the network. Figure 5 displays the network with each different color representing each of the twenty six co-authorship communities.

From Fig. 5 we can observe that only one of the communities – the one represented by yellow nodes – stands out on the network. In fact, there is a single community with 39 authors and 121 co-authorships, whereas all the others have no more than 13 authors and 28 co-authorships. Moreover, about 65% (17 in 26) of the communities have less than 10 authors (Fig. 6).

The communities provided by the ABCD algorithm, by contrast, are much more homogeneous, as it can be seen in Fig. 7 and Fig. 8: none of the communities has more than 29 authors and 37 co-authorships, and around 60% of them have no more than 10 authors.

This means that in both algorithms, 60% of the authors of the original network were discarded because they could not be putted in a community with more than four authors.

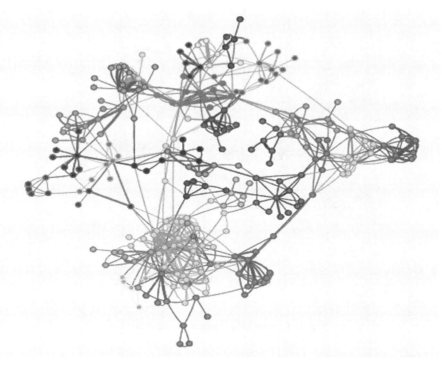

Fig. 5. Communities identified with MCL.

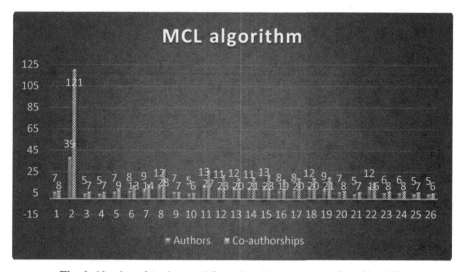

Fig. 6. Number of Authors and Co-authorships by community with MCL.

Hence, an explanation for this may be that the small number of authors in communities is due to the weak force of attractiveness of the co-authorships. Insofar as we have

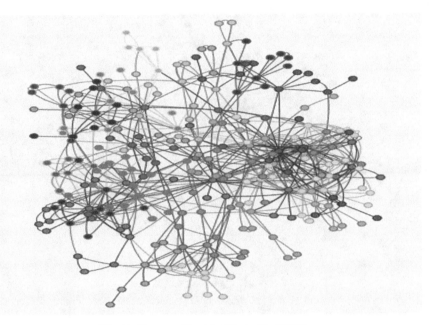

Fig. 7. Communities identified with ABCD.

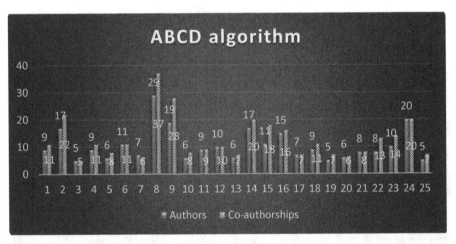

Fig. 8. Number of Authors and Co-authorships by community with ABCD.

determined that only a community with four authors is valid, it was expectable that many authors would be purged. From the 1755 co-authorships in the graph, 1225 (70%) have an attractive force equal or less than 1. Therefore, the algorithms generated small communities or even standalone authors, as solo communities.

6 Conclusions

The scientific information provided online by Ciência-IUL and the framework described were able to construct a co-authorship network and identify patterns of knowledge sharing in it. The distributed tool provides this information online and gathers important statistically information, like the total of authors, publications and co-authorships. Both the MCL and the ABCD algorithms found distinct communities using the force of Newman. However, this fact amounts to a challenge in the domain of community detection. Community detection is a formal, mathematical exercise. Once communities are automatically detected, we find a need to characterize the communities, that is, to provide the user with information in order to infer the identity or dynamics underlying the emergence of the community. In order to do so, some of the information provided may include the name, school, department and/or research center of each community member, the scientific areas of publications, the most frequent journals in the community etc. With this information the user will be able to compare the results of the different algorithms, as well as more easily confront the resulting patterns with his/her own expectations about the communities in which he/she thinks is integrated in. In this paper we describe a framework for easy and distributed access to this information which opens the doors to further modular improvements of the solution over the Web.

References

Ciência-IUL: Documentação da API Pública November 2018. https://ciencia.iscte-iul.pt/api/doc

Dongen, S.v.: Graph Clustering by Flow Simulation. University of Utrecht (2000)

Dongen, S.v.: MCL - a cluster algorithm for graphs January 2017. http://micans.org/mcl/

GEXF Working Group: GEXF File Format (2018). https://gephi.org/gexf/format/

Jackson, M.O.: Social and Economic Networks. Princeton University Press (2008)

Liu, X., Bollen, J., Nelson, M.L., Sompel, H.V.: Co-authorship networks in the digital library research. Community Inf. Process. Manage. **41**(6), 1462–1480 (2005)

MongoDB: MongoDB (2017, 2 6). https://www.mongodb.com

Newman, M.E.: Co-authorship networks and patterns of scientific collaboration. In: Proceedings of the National Academy of Sciences, pp. 5200–5205 (2004)

Newman, M.E.J.: Scientific collaboration networks. II. Shortest paths, weighted networks, and central. Phys. Rev. E **64**(1), 01632 (2001)

Node.js: Node.js (2017, 2 6). https://nodejs.org

Ruifang Liua, S.F.: Weighted graph clustering for community detection of large social networks. In: 2nd International Conference on Information Technology and Quantitative Management, ITQM (2014)

Intellectual Authorities and Hubs
of Green Chemistry

Leonardo Victor Marcelino[✉], Adilson Luiz Pinto, and Carlos Alberto Marques

Universidade Federal de Santa Catarina, R. Eng. Agronômico Andrei Cristian Ferreira s/n,
Florianópolis 88040-900, Brazil
leovmarcelino@gmail.com

Abstract. Green Chemistry (GC) is an answer to the problems ensuing from chemical pollution, adopting a proactive, prevention-based stance. After nearly three decades of its coming into being, the advances in the field towards ground-breaking chemical practice are still under discussion and new research is needed in order to systematize existing knowledge and to point to efficient means to select information. Our purpose with this study is to broaden the understanding on GC research structure, by pointing the researchers that have most contributed to its growth, spread and consolidation (its intellectual hubs), and the authors upon whose knowledge they have drawn (intellectual authorities). We analyzed 14,142 documents either containing the term "green chemistry" or published in *Green Chemistry* and *Green Chemistry Letters and Reviews* between 1990 and 2017, using network analysis and co-citation analysis. Fourteen hubs were found, and twenty-one intellectual authorities, distributed along six big specialties, previously described in the literature. Results corroborate previous analyses of the field, but this research has the advantage of stemming from the dynamics of scientific production, rather than from previously defined qualitative categories of the field itself.

Keywords: Green Chemistry · Co-citation analysis · CiteSpace

1 Introduction

There is currently a worldwide concern with preserving the environment and drastically reducing global warming. Chemistry seems to be a key agent in the current state of affairs, as it may be one of the most important sciences contributing to either aggravate or completely solve the problem. A solution within the scientific community has been growing since 1990 [1, 2], the so-called *Green Chemistry* (GC). The number of papers advocating CG-based solutions went from n = 133 (1999) to n = 3,093 (2018), a growth of more than 2,000% in two decades, as research on the Web of Science shows.

GC appears as a response to the limits of pollution and risk control policy (retroactive measures) [3], adopting a proactive, prevention-based approach. In the wake of the Pollution Prevention Act [4], GC's strategy is to reduce the use and disposal of hazardous substances, by establishing twelve principles for a green practice [5]: (1) Waste

© ICST Institute for Computer Sciences, Social Informatics and Telecommunications Engineering 2020
Published by Springer Nature Switzerland AG 2020. All Rights Reserved
R. Mugnaini (Ed.): DIONE 2020, LNICST 319, pp. 190–209, 2020.
https://doi.org/10.1007/978-3-030-50072-6_15

Prevention; (2) Atom Economy; (3) Less Hazardous Chemical Synthesis; (4) Designing Safer Chemicals; (5) Safer Solvents and Auxiliaries; (6) Design for Energy Efficiency; (7) Use of Renewable Feedstocks; (8) Reduce Derivatives; (9) Catalysis; (10) Design for Degradation; (11) Real Time Pollution Prevention; and (12) Inherently Safer Chemistry for Accident Prevention.

Almost three decades after GC was created, the advances of the field for ground-breaking chemical practice are still under discussion, usually from the perspective of its principles [2, 8–10]. Recent studies [6, 7] have detached themselves from the Twelve Principles, and sought research topics emerging from co-citation analysis of the papers published in the field, finding six main lines of research (*big specialties*) organized around sixteen smaller thematic groups, called *research specialities* (Table 1).

Table 1. Specialties in GC research. Adapted from Marcelino, Pinto and Marques [6] and Marcelino and Marques [7].

Big speciality	Research specialties
A-Solvents	#4 – Organic Reaction in Aqueous Media
	#5 – Supercritical Solvents
	#17 – Deep Eutectic Solvents
	#18 – Organic Reaction in Aqueous Media
B-Ionic Liquids	#0 – Ionic Liquids
	#6 – Recycling and Recovery of Solvents
	#8 – IL Toxicity
	#12 – IL Preparation
C-Biomass	#3 – Biomass Transformation
	#11 – Lignin Valorization
	#16 - Glycochemistry
D-Catalysis	#2 – Metal Catalysis and Microwave Activation
	#9 – Solid Acid Catalysis
	#10 – Catalytic Oxidation of Alcohols
E – GC Characterization	
F – CO_2 as Substrate	

Each specialty is conceptualized and visualized as the relationship between a time-variant network $\Phi(t)$ of co-cited papers (its intellectual base $\Omega(t)$) and the papers that promote co-citations in the network (its research front $\Psi(t)$) [11, 12], as presented in Eq. (1).

$$F(t) : Y(t) \rightarrow W(t) \qquad (1)$$

Taking into account GC's potential for an environmentally benign chemical practice in the current context of the environmental crisis, research is needed that systematizes the knowledge already produced, and points to more effective means of selecting information. The above-mentioned study [6] presents a broad overview of GC research, and provides important information on the structure and relationship between specialties, thus helping researchers in the field trace relevant information for their researches more efficiently. Our purpose with this research is to broaden the understanding on GC research structure, by pointing the researchers that have most contributed to its growth, spread and consolidation (its intellectual hubs), and the authors upon whose knowledge they have drawn (intellectual authorities).

Section 2 discusses the concepts of *hubs* and *intellectual authorities* in the context of the specialties defined around an intellectual base and a research front. Section 3 details methodology. Section 4 presents the hubs and intellectual authorities for each of GC's big specialties (4.1 Intellectual Hubs and Authorities of GC Specialties) and validates the data based on prior discussions in the literature (4.3 Results validation). Finally, we present the conclusions and references. The complete table with the research data can be accessed in the Supplementary Material[1].

2 Hubs and Intellectual Authorities

Jon Kleinberg [13] has developed an algorithm to analyze the importance of network nodes in providing content of relevance (authorities) or the ability to select, group and disseminate information (hubs). Kleinberg's algorithm is based on mutual recursion, so that, in a given network, an important hub connects to important authorities, and vice versa.

In the case of the present study, we have a structured co-citation network with an intellectual base (cited papers) and information on the research fronts, the papers that cite the works of this network [11, 12]. This would mean that there is an invisible network formed by the research front that interacts with the visible network made by the intellectual base. By considering these two sources of information, we have made Kleinberg's analysis of hubs and authorities more complex, and developed an alternative thereto.

We consider that certain authors within the research front act *as* hubs: nodes with high degree of connection, grouping the intellectual base around a research theme and disseminating information throughout the scientific community. We have two indicators of hub behavior of a research front paper: its Coverage (Cov) value and Global Citation Score (GCS). Coverage gives us the concentration intensity, how much a certain author contributes to grouping information and shaping a research/specialty front. It is a measure associated with the author's contribution to tie research around a specific theme. The GCS is the amount of citations a given paper has received within the scientific community, and provides an idea of the role of the researcher in provoking the diffusion of the research/specialty front or how much it has been disseminated within the Chemistry community.

Some authors of the intellectual base act *as* authorities: nodes with relevant information that are linked by several dissemination nodes, the hubs. Intellectual authorities

---

[1] http://bit.do/ESM_husbs-authorities.

provide the concepts, methodologies and techniques necessary for the existence of a specialty. There are three indicators defined by Chen [11, 12] that we used for identifying intellectual authorities: citation frequency (CF), citation burstness (CB), and betweenness centrality (BtC). CF measures how many times the author was cited within the selected sample, which can be called the *Local Citation Score* (LCS). Unlike GCS, which considers all citations of a given piece of work in all the literature, CF is an indicator of the recognition of the paper/author by green chemists (a group of previously selected works), which allows us to draw inferences on the recognition of a given specialty. CB measures a sudden increase in the number of citations to a document/author, and is an indicator of the degree of innovation introduced in the field of knowledge, which generates sudden interest in a given subject. BtC is the ability of a work/author to transit among various specialties, promoting interaction and information exchange. These elements with high BtC fill gaps in the structure of knowledge and act as bridges among different research topics.

To find these hubs and authorities, we consider several metrics, adding the values of each author's output to reach a total. For a given author, LCS, number of papers (NP), CF, BtC and CB values are all calculated by the simple sum of the individual values of each metric for an author's papers. Total GCS is the sum of GCS values of different papers by an author. Nevertheless, a given authors' total coverage is measured by the quotient of the number of papers this author cites in the intellectual base by the total number of papers in the intellectual base of a specialty.

3 Methodology

All data were collected in May 2018 in the Web of Science database (Core Collection, indexes: SCI-EXPANDED, SSCI, A&HCI, CPCI-S, CPCI-SSH, ESCI) from 1990 (the decade in which GC emerges [1, 2]) to 2017 (he last full year at the time of analysis). We selected documents containing references, such as: papers, reviews, proceedings papers and book chapters (henceforth *papers*). We searched for the term "green chemistry" in titles, keywords and abstracts, thus indicating explicit affiliation to GC. We also selected all texts published in *Green Chemistry Journal* and *Green Chemistry Letters and Reviews*, as these are GC journals, and their texts do not always use the descriptor "green chemistry" in their titles, abstracts and keywords.

We obtained a final set of 14,142 different records (duplicates excluded), being 8,586 records with the term "green chemistry" and 6,004 in specialized journals. We analyzed this data using CiteSpace (version 5.2.R1.3.9.2018), considering: Look Back Year (LBY): -1; Time Slice: 3 years (1990–2017); Node types: Cited reference; Top N.: 100. These parameters were chosen for their best results in Modularity Q and Silhouette values. Other parameters followed default program settings. After the network was created, the algorithm was applied for the generation of clusters, the themes of which were inferred from the titles and abstracts of the papers in the research front of each cluster. The authors names were standardize to avoid duplicates, specially the given names that were fully abbreviated without punctuations and spaces.

The validation of the analysis is made by considering the internal evaluation parameters of the network (Modularity Q and silhouette) and peer recognition (comparison with previous qualitative description of the field made by renowned researchers) [2, 14], important criteria for validation, as Noyons [15] points out.

4 Results and Discussion

4.1 Intellectual Hubs and Authorities of GC Specialties

A – Solvents. The analysis of the authors who contribute the most to the dissemination and consolidation of a specialty (its intellectual hubs) may be done by means of the coverage value that they accumulate in a big specialty, that is, by the proportion of papers co-cited by a text/author, given the total amount of papers cited in the intellectual base of the specialty or big especialty (BSp).

Big specialty A – Solvents has 43 papers in its research front, among which Victorio Cadierno and Javier Franco have the highest coverage values (0.24 and 0.18 respectively) and also the highest number of papers (Table 2). The next ten authors are all related to the same paper in the research front of small specialty #5 – Supercritical Solvents. As this front is composed of only two papers, its coverage is high. Pascale Crochet occupies the third position in number of papers, with 3 texts in A – Solvents, and an accumulated coverage of 0.14. Paul T. Anastas, Nicolas Eghbali, Vivek Polshettiwar and Rajender S. Varma possess the highest citation values, but do not have high coverage values, meaning that they have papers acknowledged by the scientific community as a whole, but they are not necessarily the ones most closely related with A – Solvents. Cadierno, Franco and Crochet, however, have a GCS of approximately 200, and follow the highest-ranking authors in this indicator.

With regard to the intellectual authorities present in the intellectual base (72 papers) of big specialty A – Solvents (Table 3), Chao-Jun Li has the largest number of papers used in the references of the research front, with a high accumulated CF of 951, almost three times as big as the runner-up (Varma, 85). Li stands out in terms of CB, with twice as many citations (117) as the second place (Abbott, 46), meaning that his works were especially important for this specialty, approaching subjects that aroused the interest of its researchers. CB for the output of this intellectual base occurs, on average, between 2006 and 2009, which is neither as old as the average CB of B – Ionic Liquids (2003–2006), nor as new as the average CB of C – Biomass (2011–2014).

Li also has the highest accumulated BtC, meaning that his works in this specialty possibly interact with other specialties, strengthening the field of GC as a whole. It should be pointed out that both A – Solvents and C – Biomass are big specialities with the lowest accumulated BtC values per author, meaning that, even though their authorities promote interaction throughout GC, their reach is more limited. Mark J. Burk e William D. McGhee have a high number of papers on the intellectual base, but these have low CF or CB, which indicates that they did not generate much interest in the community, even if they are important for this big specialty.

We may, therefore, conclude that Victorio Cadierno and Javier Franco are the leading authors in terms of output and contributions to co-citations in the research fronts of big specialty A – Solvents, and may be considered its intellectual hubs. Chao-Jun Li is the author with the highest number of papers in the intellectual base, accumulating the highest CF, CB and BtC, demonstrating his importance in providing support to the works

Table 2. Big specialty A – Solvents, most relevant authors of the research front.

Author	Cov A	Total Cov	GCS A	Total GCS	LCS A	Total LCS	N BSp
CADIERNO, V	0.2411	0.0527	245	397	5	7	2
FRANCOS, J	0.1851	0.0417	200	352	4	6	2
BORKOWSKY, SL	0.1667	0.0327	79	79	1	1	1
BROWN, GH	0.1667	0.0327	79	79	1	1	1
BURK, MJ	0.1667	0.0327	79	79	1	1	1
FENG, S	0.1667	0.0327	79	79	1	1	1
GROSS, MF	0.1667	0.0327	79	79	1	1	1
LELACHEUR, RM	0.1667	0.0327	79	79	1	1	1
LUAN, L	0.1667	0.0327	79	79	1	1	1
MORGENSTERN, DA	0.1667	0.0327	79	79	1	1	1
MORITA, DK	0.1667	0.0327	79	79	1	1	1
TUMAS, W	0.1667	0.0327	79	79	1	1	1
CROCHET, P	0.1421	0.0409	197	393	3	7	4
JIANG, B	0.0990	0.0194	175	175	2	2	1
LI, GG	0.0990	0.0194	175	175	2	2	1

in big specialty A – Solvents, acting as an intellectual authority for these studies. Within this big specialty, it is important to highlight the importance of the research in organic reactions in aqueous medium (cluster #18), as it is one of its most recognized subjects of study, and has the largest research front.

B – Ionic Liquids. This big specialty has the second smallest research front, with 27 papers published between 1996 and 2017. Table 4 presents the most important authors of the research fronts of big specialty B – Ionic Liquids, classified in descending order of coverage. Five authors stand out with values above 0.2: Robin D. Rogers, John D. Holbrey, W. Matthew Reichert, Richard P. Swatloski and Kenneth R. Seddon. All five of them have a large number of papers concentrated in the research front of this big specialty, in particular Rogers and Seddon, with four papers each. Seddon also stands out in the number of citations accumulated by his papers, with 1,240, followed by Louis C. Branco, João N. Rosa, Joaquim J. M. Ramos e Carlos A. M. Alfonso, who accumulated 948 citations of their papers.

The intellectual base of this big specialty has 112 papers, the largest of them all, with papers published between 1973 and 2011. Due to the size of the intellectual base, there is a large number of papers published by the authors (Table 5): Kenneth R. Seddon has 17 papers, all within this great specialty, showing his dedication and contribution to the theme. Other authors with a large number of publications include Robin D. Rogers, John D. Holbrey, Tomas Welton, Richard P. Swatloski and Peter Wasserscheid. Four of these authors stand out due to the high CF and CB accumulated by their works, demonstrating the interest that their studies have raised for research on ionic liquids: Welton has the

Table 3. Big specialty A – Solvents, most relevant authors of the intellectual base.

Authors	NP A	Total NP	N BSp	CB A	Total CB	BtC A	Total BtC	CF A	Total CF
LI, CJ	10	9	2	117.46	133.09	0.17	0.2	951	1074
VARMA, RS	9	2	2	32.11	118.59	0.05	0.21	85	426
JESSOP, PG	6	2	4	12.04	89.5	0.09	0.32	20	279
BURK, MJ	4	4	1	5.25	5.25	0.09	0.09	14	14
MCGHEE, WD	4	4	1	0	0	0.01	0.01	9	9
DESIMONE, JM	4	3	2	4.59	18.79	0.01	0.01	11	107
CHAN, TH	3	3	1	44.42	44.42	0.02	0.02	194	194
IKARIYA, T	3	2	2	12.04	33.2	0.09	0.13	20	57
NOYORI, R	3	2	2	12.04	33.2	0.09	0.13	20	57
JEROME, F	3	1	3	28.21	63.23	0.01	0.2	87	180
TUMAS, W	3	1	3	5.25	21.7	0.09	0.09	8	108
BRESLOW, R	2	2	1	19.69	19.69	0.06	0.06	89	89
ABBOTT, AP	2	2	1	46.39	46.39	0.01	0.01	128	128
DOMLING, A	2	2	1	19.66	19.66	0.02	0.02	216	216
LINDSTROM, UM	2	2	1	31.32	31.32	0.06	0.06	332	332
BIENAYMÉ, H	2	2	1	20.42	20.42	0.03	0.03	236	236
FOKIN, VV	2	2	1	24.08	24.08	0.02	0.02	358	358
FINN, MG	1	1	1	3.99	3.99	0.01	0.01	190	190
KOLB, HC	1	1	1	3.99	3.99	0.01	0.01	190	190
MULDOON, J	1	1	1	3.99	3.99	0.01	0.01	190	190
SIMON, MO	1	1	1	26.61	26.61	0.07	0.07	140	140
ROYER, S	1	1	1	28.21	28.21	0.01	0.01	87	87
VIGIER, KD	1	1	1	28.21	28.21	0.01	0.01	87	87
ZHANG, QH	1	1	1	28.21	28.21	0.01	0.01	87	87
FENG, SG	1	1	1	5.25	5.25	0.09	0.09	8	8
GROSS, MF	1	1	1	5.25	5.25	0.09	0.09	8	8
CHEN, L	1	1	1	9.16	9.16	0.02	0.02	192	192
BLACKERT, JF	1	1	1	4.01	4.01	0.1	0.1	6	6
TANKO, JM	1	1	1	4.01	4.01	0.1	0.1	6	6

highest accumulated CF value (831), followed by Seddon (774), Wasserscheid (463) and Rogers (335). Seddon, however, presents the highest CB value (2192.7), followed by

Table 4. Big specialty B – Ionic Liquids, most relevant authors of the research front.

Author	Cov B	Total Cov	GCS B	Total GCS	LCS B	Total LCS	N BSp
ROGERS, RD	0.295	0.090	612	612	4	4	1
HOLBREY, JD	0.231	0.070	584	584	3	3	1
REICHERT, WM	0.221	0.067	594	594	3	3	1
SWATLOSKI, RP	0.221	0.067	594	594	3	3	1
SEDDON, KR	0.201	0.061	1240	1240	4	4	1
AFONSO, CAM	0.161	0.049	948	948	2	2	1
BRANCO, LC	0.161	0.049	948	948	2	2	1
RAMOS, JJM	0.161	0.049	948	948	2	2	1
ROSA, JN	0.161	0.049	948	948	2	2	1
BROKER, GA	0.157	0.048	566	566	2	2	1
KITAZUME, T	0.142	0.046	164	187	3	4	2
VISSER, AE	0.142	0.043	169	169	2	2	1
ZULFIQAR, F	0.117	0.036	143	143	2	2	1
STOLTE, S	0.090	0.027	200	200	2	2	1
THOMING, J	0.090	0.027	200	200	2	2	1

Holbrey, Rogers and Welton with aproximately 1,000 each. Among these four authors, Seddon stands out for his accumulated BtC value (0.31), showing that his works allow for interaction among specialties.

Considering the number of papers on the research fronts and the high number of accumulated citations, we can say that Robin D. Rogers and Kenneth R. Seddon are the main intellectual hubs of big specialty B – Ionic Liquids, with great contribution to set the boundaries and spread discussions of this specialty. Other authors displaying hub behavior are John D. Holbrey, W. Matthew Reichert and Richard P. Swatloski. In the intellectual base, Kenneth R. Seddon also has great importance, as he is the most published and has the highest accumulated CB value, and second highest CF value, acting also as intellectual authority of this great specialty. Other revelant authors as intellectual authorities on ionic liquids are Robin D. Rogers, Tom Welton and Peter Wasserscheid. The recurrence of authors in the research front and intellectual base demonstrates the internal cohesion of this big specialty; additionally, the fact that this big specialty holds the highest CB values indicates the great interest and commitment that the theme has aroused among its researchers, especially until 2009.

C – Biomass. This research front is the fourth largest within GC, with 42 papers published from 1999 to 2017. The authors with the highest coverage in this specialty are David M. Alonso, James A. Dumesic and Atsushi Takagaki, all with 0.202 each, and two appearances on the research fronts (Table 6). Michikazu Hara comes third in accumulated coverage, and appears four times in the research fronts, the highest LCS. Yan-Long Gu has the second highest LCS value, appearing three times in the research fronts, despite his

Table 5. Big specialty B – Ionic Liquids, most relevant authors of the intellectual base.

Authors	NP B	Total NP	N BSp	CB B	Total CB	BtC B	Total BtC	CF B	Total CF
SHELDON, R	23	4	5	92,15	275,84	0,06	0,5	313	1718
SEDDON, KR	17	17	1	228,41	228,41	0,31	0,31	774	774
ROGERS, RD	12	9	3	113,43	149,52	0,12	0,27	335	612
SWATLOSKI, RP	9	7	2	76,4	94,37	0,07	0,22	141	359
HOLBREY, JD	8	7	2	107.89	114.85	0,07	0,21	206	403
WELTON, T	8	7	2	136,34	152,53	0,18	0,18	831	869
JASTORFF, B	7	7	1	81,96	81,96	0,15	0,15	182	182
RANKE, J	7	7	1	81,96	81,96	0,15	0,15	182	182
VISSER, AE	6	6	1	84,45	84,45	0,11	0,11	190	190
EARLE, MJ	6	6	1	66,77	66,77	0,03	0,03	197	197
WASSERSCHEID, P	6	6	1	114.14	114.14	0,03	0,03	463	463
DE SOUZA, RF	5	5	1	58.69	58.69	0.16	0.16	288	288
DUPONT, J	5	5	1	58.69	58.69	0.16	0.16	288	288
SUAREZ, PAZ	4	4	1	58.69	58.69	0.16	0.16	284	284
DULLIUS, JEL	3	3	1	27.18	27.18	0.16	0.16	60	60
EINLOFT, S	3	3	1	27.18	27.18	0.16	0.16	60	60
KEIM, W	1	1	1	41.22	41.22	0,02	0,02	311	311
KLEMET, I	1	1	1	3.96	3.96	0.25	0.25	6	6
KNOCHEL, P	1	1	1	3.96	3.96	0.25	0.25	6	6
LUTJENS, H	1	1	1	3.96	3.96	0.25	0.25	6	6

ranking 7th in accumulated coverage, with 0.162. Alonso and Dumesic have a high GSC value, being cited 1,061 times each, but it is Joseph J. Bozell and Gene R. Petersen, who hold the highest accumulated CF value, 1,404 each, thus establishing their importance for biomass research.

The intellectual base of C – Biomass contains 49 works, published between 1993 and 2016. James A. Dumesic stands out with the highest number of papers (9), followed by George W. Huber (5) (Table 7). The high number of contributions to C – Biomass indicates that they their output grounds the discussions on these specialties. The authors with the highest accumulated CF are Avelino Corma and Sara Iborra with 717 each, followed by James A. Dumesic (661) and George W. Huber (556); these authors are the most acknowledged by the scientific community on this intellectual base. The same four authors are the ones with the highest CB value, which represents how their output has arisen the interest and commitment of research in this specialty. Dumesic's accumulated CB (1,035.5) is worthy of note, being almost twice as big as that of his runner-up

Table 6. Big specialty C – Biomass, most relevant authors of the research front.

Author	Cov C	Total Cov	GCS C	Total GCS	LCS C	Total LCS	N BSp
ALONSO, DM	0.202	0.027	1061	1061	2	2	1
DUMESIC, JA	0.202	0.027	1061	1061	2	2	1
TAKAGAKI, A	0.202	0.038	204	230	2	3	2
HARA, M	0.182	0.071	291	485	4	7	2
RENDERS, T	0.163	0.022	14	14	2	2	1
SELS, BF	0.163	0.030	14	122	2	3	2
GU, YL	0.162	0.029	378	707	3	5	3
HAYASHI, S	0.141	0.049	160	249	2	4	2
BELL, AT	0.141	0.019	233	233	1	1	1
CHIDAMBARAM, M	0.141	0.019	233	233	1	1	1
CLIMENT, MJ	0.141	0.019	413	413	1	1	1
IBORRA, S	0.141	0.019	413	413	1	1	1
CORMA, A	0.141	0.022	413	445	1	2	2
GALKIN, MV	0.122	0.016	1	1	1	1	1
KUMANIAEV, I	0.122	0.016	1	1	1	1	1

(Corma, 672,9). The average CB of C – Biomass spans from 2011 to 2014, the most recent interval, which may indicate the interest that the most recent studies have aroused in this topic and its relevance. With respect to BtC values, the most relevant authors are Robin D. Rogers and Richard P. Swatloski, both highly representative for big specialty B – Ionic Liquids, indicating their contribution to promote interaction between these two research topics.

Atsushi Takagaki, Michikazu Hara, David M. Alonso and James Dumesic can be considered the intellectual hubs of the big specialty C – Biomass, due to their high coverage and their number of papers on the research fronts, indicating their high contribution to set work themes and spread research on biomass. Other important authors are Joseph J. Bozell and Gene R. Peterson, whose work has high circulation among the scientific community. James A. Dumesic has a very important role as intellectual authority in this specialty, as he is the author who publishes the most, and whose texts have aroused the interest of the scientific community (high CB and CF). Other authors who also play an important role as intellectual authorities are George W. Huber, Avelino Corma and Sara Iborra.

D – Catalysis. This big specialty has the second largest research front, with 46 papers published between 1999 and 2014. Michikazu Hara has the highest coverage and is also high-ranking in number of appearances on the research front (LCS) (Table 8). Xuezheng Liang, Chenze Qi and Daizo Yamaguchi follow Hara on coverage ranking, but have low

Table 7. Big specialty C – Biomass, most relevant authors of the intellectual base.

Authors	NP C	Total NP	N BSp	CB C	Total CB	BtC C	Total BtC	CF C	Total CF
DUMESIC, JA	9	9	1	113.47	113.47	0.04	0.04	661	661
HUBER, GW	5	5	1	66.23	66.23	0.01	0.01	556	556
KOBAYASHI, S	4	5	2	8.58	16.72	0.1	0.1	22	73
ALONSO, DM	4	4	1	60.87	60.87	0.04	0.04	221	221
CHHEDA, JN	4	4	1	39.5	39.5	0	0	333	333
CORMA, A	3	5	2	67.29	92.1	0.08	0.08	717	772
HACHIYA, I	3	3	1	3.96	3.96	0.04	0.04	15	15
IBORRA, S	3	3	1	67.29	67.29	0.08	0.08	717	717
ROMAN-LESHKOV, Y	3	3	1	30.25	30.25	0	0	219	219
ROGERS, RD	2	12	3	17.97	149.52	0.15	0.27	218	612
SWATLOSKI, RP	2	9	2	17.97	94.37	0.15	0.22	218	359
BOND, JQ	2	2	1	39.97	39.97	0.04	0.04	155	155
GALLEZOT, P	2	2	1	62.15	62.15	0	0	179	179
DAVISON, BH	2	2	1	36.25	36.25	0.15	0.15	241	241
RAGAUSKAS, AJ	2	2	1	36.25	36.25	0.15	0.15	241	241
TSCHAPLINSKI, T	2	2	1	36.25	36.25	0.15	0.15	241	241
HOLBREY, JD	1	8	2	6.96	114.85	0.14	0.21	197	403
SPEAR, SK	1	2	2	6.96	25.08	0.14	0.14	197	256
VELTY, A	1	1	1	19.77	19.77	0.07	0.07	381	381
BRITOVSEK, G	1	1	1	6.42	6.42	0.13	0.13	165	165
CAIRNEY, J	1	1	1	6.42	6.42	0.13	0.13	165	165

GCS, indicating that their work does not have much circulation within the academic community. Rajender S. Varma is a prominent author within this specialty, having the highest number of citations, the highest number of papers and high coverage.

The intellectual base of this big specialty consists of 61 papers, published between 1981 and 2017. Rajender S. Varma, Roger Sheldon, Paul T. Anastas and André Loupy are the most recurring authors (Table 9). Sheldon has a large number of papers in this intellectual base, but also in other major specialties, showing that his research is not as specialized, whereas Anastas has 6 appearances (out of 9) in big specialty D – Catalysis, highest CB value, as well as the largest highest number of accumulated citations and BtC, showing his expertise in this field. Tracy C. Williamson, Mary M. Kirchhoff and Nicolas Eghbali are co-authors with Anastas; they appear among high-ranking authors in CF, CB and BtC. Varma's influence is corroborated by his high CF, CB and BtC

Table 8. Big specialty D – Catalysis, most relevant authors of the research front.

Author	Cov D	Total Cov	GCS D	Total GCS	LCS D	Total LCS	N BSp
HARA, M	0.279	0.0707	194	485	3	7	2
LIANG, XZ	0.264	0.0438	22	22	3	3	1
QI, CZ	0.264	0.0438	22	22	3	3	1
YAMAGUCHI, D	0.197	0.0381	63	126	2	4	2
VARMA, RS	0.181	0.0434	2060	3938	4	6	3
HAYASHI, S	0.179	0.0486	89	249	2	4	2
GUO, Y	0.130	0.0216	134	134	2	2	1
KATO, H	0.114	0.0217	63	126	1	2	2
KITANO, M	0.114	0.0217	63	126	1	2	2
NAKAJIMA, K	0.114	0.0217	63	126	1	2	2
SUGANUMA, S	0.114	0.0217	63	126	1	2	2
GUO, YX	0.098	0.0163	14	14	1	1	1
XIAO, HQ	0.098	0.0163	14	14	1	1	1
HU, BW	0.083	0.0137	2	2	1	1	1
LI, CQ	0.083	0.0137	2	2	1	1	1

values. Loupy and Polshettiwar also have high CB, indicating that they have developed research that has leveraged investigations into this major specialty at certain periods.

In analyzing the authors of the research front, Rajender S. Varma and Michikazu Hara may be considered to be the most important intellectual hubs, largely responsible for promoting the cohesion of this big specialty around common research topics. Varma is also a strong intellectual authority, as he has a large number of papers, and high CF and CB. Other important authors are André Loupy and Vivek Polshettiwar, with large numbers of papers, and high CB and accumulated CF. Paul T. Anastas is greatly influential in the specialty, despite his output consisting of more general papers on GC.

E – GC Characterization. This specialty is formed by a single large cluster that has the largest research front (even when compared to the big specialties), with 56 published papers between 1999 and 2017. This specialty is plural in its themes, grouping papers concerned with laying the foundations, challenges and possibilities of GC, as well as setting its themes. The authors with the highest accumulated coverage are Paul T. Anastas, Roger A. Sheldon and James H. Clark, all of whom are highly known in the field, which can be seen by the high number of citations they receive (Table 10).

With regard to the intellectual base, it contains 45 works (the 5th largest if compared to the big specialties), published between 1991 and 2016. Roger A. Sheldon is the author with the highest number of works in the intellectual base of this specialty, although a large number of his papers is to be found in other specialties, thus showing he has more transversal influence throughout GC (Table 11). James H. Clark, David J. C. Constable, Alan D. Curzons and Barry M. Trost have a large number of papers with great focus on

Table 9. Big specialty D – Catalysis, most relevant authors of the intellectual base.

Authors	NP D	Total NP	N BSp	CB D	Total CB	BtC D	Total BtC	CF D	Total CF
VARMA, RS	7	9	2	86.48	118.59	0.16	0.21	341	426
SHELDON, R	6	23	5	26.64	275.84	0.02	0.5	221	1718
ANASTAS, P	6	9	2	85.36	115.73	0.28	0.46	1065	2410
LOUPY, A	5	5	1	70.51	70.51	0.02	0.02	148	148
POLSHETTIWAR, V	3	4	2	55.41	66.15	0.03	0.03	180	199
DOMEN, K	3	3	1	46.33	46.33	0.09	0.09	139	139
HARA, M	3	3	1	46.33	46.33	0.09	0.09	139	139
HAYASHI, S	3	3	1	46.33	46.33	0.09	0.09	139	139
KONDO, JN	3	3	1	46.33	46.33	0.09	0.09	139	139
ARENDS, I	3	3	1	15.54	15.54	0.01	0.01	180	180
WILLIAMSON, TC	2	3	2	35.93	50.41	0.09	0.27	116	1229
PETIT, A	2	2	1	32.71	32.71	0	0	57	57
TANAKA, K	2	2	1	19.82	19.82	0.1	0.1	268	268
KIRCHHOFF, MM	2	2	1	18.53	18.53	0.15	0.15	532	532
POLIAKOFF, M	1	6	5	0	41.82	0	0.02	182	354
METZGER, JO	1	3	2	8.15	19.88	0.11	0.13	14	33
TODA, F	1	2	2	0	0	0.08	0.11	220	225
EGHBALI, N	1	1	1	32.89	32.89	0.07	0.07	238	238
FARREN, TR	1	1	1	0	0	0	0	182	182
FITZPATRICK, JM	1	1	1	0	0	0	0	182	182

this specialty. These authors also have high CF, CB and BtC values, corroborating their strong role in substantiating the characterization of GC. Concepción Jiménez-González has a high CB value, showing that her ideas have aroused great interest. Philip G. Jessop accumulates great BtC value, indicating his role in promoting interaction between this specialty and the others. Average CB of this specialty spanned from 2007 to 2011.

As pointed above, most outstanding authors on the research front in this specialty are Paul T. Anastas, Roger A. Sheldon and James H. Clark, indicating that they are intellectual hubs, responsible for bringing together the discussion about what GC is and how to evaluate it, as well as in spreading the principles and procedures of this field. The intellectual base of this specialty shows us that Sheldon and Clark are also important intellectual authorities, developing research that provides important information for GC characterization. David J. C. Constable, Alan D. Curzons and Barry M. Trost also play an important role as intellectual authorities in the characterization of GC. We shall later

Table 10. Specialty E – GC Characterization, most relevant authors of the research front.

Author	Cov E	Total Cov	GCS E	Total GCS	LCS E	Total LSC	N BSp
ANASTAS, P	0.19	0.071	1561	5198	3	9	5
SHELDON, RA	0.13	0.049	433	1402	2	4	3
CLARK, JH	0.13	0.016	625	625	1	1	1
LAWRENSON, S	0.11	0.013	18	18	2	2	1
NORTH, M	0.11	0.022	18	595	2	3	2
VACCARO, L	0.11	0.013	56	56	2	2	1
NAMIESNIK, J	0.11	0.022	93	180	2	3	2
PENA-PEREIRA, F	0.11	0.022	93	180	2	3	2
EGHBALI, N	0.11	0.039	891	4455	1	5	5
JEROME, F	0.09	0.027	279	575	2	4	3
ACKERMANN, L	0.09	0.011	16	16	1	1	1
BAO, WL	0.09	0.011	10	10	1	1	1
BOUSFIELD, TW	0.09	0.011	4	4	1	1	1
BUCO, A	0.09	0.011	1	1	1	1	1
CAMP, JE	0.09	0.011	4	4	1	1	1

discuss the existence of hubs and intellectual authorities for the whole field of GC, extrapolating the limits of specialist expertise.

F – CO_2 as substrate. Specialty F – CO_2 as Substrate has the smallest research front (18 works) and the smallest intellectual base (28); although it is thematically close to C – Biomass, it presents a co-citation network with very distinct characteristics, forming an isolated specialty. In their research front, Johannes Steinbauer and Thomas Werner are notable for their high coverage (0.47 each) and number of papers (4 each), which are co-authored by Hendrik Büttner, Lars Longwitz and Christoph Wulf (Table 12).

Within the intellectual base of the specialty F – CO2 as Substrate, we can highlight the contributions of Toshiyasu Sakakura and his co-author, Jun-Chul Choi, with a high number of papers, CF and accumulated BtC (Table 13). Philip G. Jessop and Walter Leitner published papers in collaboration, and also have high number of publications, highest accumulated CB and BtC in this specialty. Average CB spanned from 2005 to 2008, a very early topic of interest.

The above shows that Johannes Steinbauer and Thomas Werner are intellectual hubs of the specialty F – CO2 as Substrate, due to their high coverage and number of papers on the research front. Intellectual authorities of this specialty may be Toshiyasu Sakakura, Jun-Chul Choi, Philip G. Jessop and Walter Leitner, for the number of papers, CF and accumulated CB values on the intellectual base.

Table 11. Specialty E – GC Characterization, most relevant authors of the intellectual base.

Authors	NP E	Total NP	N BSp	CB E	Total CB	BtC E	Total BtC	CF E	Total CF
SHELDON, R	9	23	5	108.85	275,84	0.38	0,5	1053	1718
CLARK, J	6	8	3	14.52	48.86	0,05	0,05	253	384
TROST, BM	4	4	1	32.55	32.55	0.3	0.3	737	737
CONSTABLE, DJC	4	4	1	27.43	27.43	0,12	0,12	475	475
CURZONS, AD	4	4	1	27.43	27.43	0,12	0,12	475	475
ANASTAS, P	3	9	2	30.37	115.73	0,18	0.46	1345	2410
DUNN, PJ	3	3	1	30.23	30.23	0,05	0,05	303	303
PRAT, D	3	3	1	0	0	0.04	0.04	59	59
HORVATH, IT	2	6	3	18.72	77.42	0	0,03	170	306
MACQUARRIE, DJ	2	3	2	10.56	19.87	0	0	16	32
JIMENEZ-GONZALEZ, C	2	2	1	45.59	45.59	0.08	0.08	222	222
CUNNINGHAM, VL	2	2	1	0	0	0,1	0,1	316	316
JESSOP, PG	1	6	4	21.26	89,5	0,15	0,32	123	279
WILLIAMSON, TC	1	3	2	14.48	50.41	0,18	0,27	1113	1229
JEROME, F	1	3	3	19.42	63,23	0,13	0.2	60	180
GU, YL	1	2	2	19.42	35.02	0,13	0.19	60	93
BROXTERMAN, QB	1	1	1	23	23	0,06	0,06	71	71
MANLEY, JB	1	1	1	23	23	0,06	0,06	71	71
PONDER, CS	1	1	1	23	23	0,06	0,06	71	71

4.2 Results Validation

In 2016, the *Green Chemistry Journal* published a series of 13 editorials commemorating the 25 years of GC, penned by renowned authors of the field, showcasing the advances made in each of the Twelve Principles. By crossing the authors of these editorials and their references with the results of this research, it is possible to provide indications of the relevance of the results found. Two out of the 14 different intellectual hubs wrote papers for the commemorative editorials, and two are listed in their references, so that 29% of the hubs are related to these editorials. Anastas, an important GC hub, writes with Han, Leitner and Poliakoff [2] the opening editorial of the commemorative series, and also appears as a reference in almost all editorials, being related to all principles except 12 – Inherently Safer Chemistry [25]. Sheldon (GC hub and authority) is the author of the editorial on Principle 2 – Atom economy [18], and appears as a reference in six editorials [19–24]. Hara (hub of big specialties C – Biomass and D – Catalysis) appears in the editorial references of Delidovich and Palkovits [19] on Principle 9 – Catalysis. On the other hand, Clark (hub and authority of specialty E – GC Characterization), appears on

Table 12. Specialty F – CO_2 as substrate, most relevant authors of the research front.

Author	Cov F	Total Cov	GCS F	Total GCS	LCS F	Total LCS	N BSp
STEINBAUER, J	0.47	0.0432	34	53	4	7	2
WERNER, T	0.47	0.0432	34	53	4	7	2
BUTTNER, H	0.25	0.0215	23	31	2	3	2
LONGWITZ, L	0.25	0.0240	10	20	2	4	2
WULF, C	0.25	0.0215	23	31	2	3	2
CANELLAS, S	0.14	0.0107	1	1	1	1	1
JOSE, T	0.14	0.0107	1	1	1	1	1
KLEIJ, AW	0.14	0.0107	1	1	1	1	1
PERICAS, MA	0.14	0.0107	1	1	1	1	1
GARCIA, H	0.11	0.0084	102	102	2	2	1
CAO, CY	0.11	0.0084	3	3	1	1	1
CHOI, JC	0.11	0.0084	1	1	1	1	1
DINDAROGLU, M	0.11	0.0084	15	15	1	1	1
FUKAYA, N	0.11	0.0084	1	1	1	1	1
Gao, XT	0.11	0.0084	2	2	1	1	1

the editorials on Principles 2 – Atom Economy [18], 9 – Catalysis [19] and 8 – Reduce Derivatives [20].

There are 21 different intellectual authorities, twelve of which are related to the editorials (57%). Chao-Jun Li is an authority of big specialty A – Solvents, and author of the editorial on Principle 8 – Reduce Derivatives. Corma and Iborra (authorities of C – Biomass) feature in the references to the editorial on Principles 9 – Catalysis [19] and 7 – Use of Renewable Feedstocks [21]. Constable and Curzons (authorities of E – GC Characterization) are featured in the references to the editorials on Principles 8 – Reduce Derivatives [20], 6 – Design for Energy Efficiency [22], and 12 – Inherently Safer Chemistry [25]. Trost, also authority in E – GC Characterization, appears in the references to the editorials on Principles 8 – Reduce Derivatives [20] and 2 – Atom Economy [18]. Jessop is an important intellectual authority in F – CO_2 as Substrate, and appears as author to the editorial on Principles 5 – Safer Solvents and Auxiliaries [26] and as a reference in 8 – Reduce Derivatives [20]. Leitner, also authority in F – CO_2 as Substrate, is present in the references to the editorials on Principles 3 – Less Hazardous Chemical Synthesis [24], 6 – Design for energy efficiency [22] and 8 – Reduce Derivatives [20]; he is also author of the opening editorial [2] Toshiyasu Sakakura and Jun-Chul Choi, both authorities in F – CO_2 as Substrate, are references in editorial about 3 – Less Hazardous Chemical Synthesis [24].

We may also note that many of the intellectual hubs are not mentioned in the commemorative editorials. This may be due to the fact that they are authors more closely

Table 13. Specialty F – CO_2 as Substrate, most relevant authors of the intellectual base.

Authors	NP F	Total NP	N BSp	CB F	Total CB	BtC F	Total BtC	CF F	Total CF
SAKAKURA, T	3	3	1	25.05	25.05	0.04	0.04	229	229
JESSOP, PG	2	6	4	44.1	89.5	0.07	0.32	116	279
LEITNER, W	2	4	3	39.04	48.93	0.09	0.1	118	134
CHOI, JC	2	2	1	12.18	12.18	0.04	0.04	167	167
ROGERS, RD	1	12	3	18.12	149.52	0	0.27	59	612
CLARK, J	1	8	3	9.31	48.86	0	0.05	16	384
POLIAKOFF, M	1	6	5	7.57	41.82	0	0.02	13	354
BECKMAN, EJ	1	4	3	18.17	36.39	0	0.02	44	87
IKARIYA, T	1	3	2	21.16	33.2	0.04	0.13	37	57
NOYORI, R	1	3	2	21.16	33.2	0.04	0.13	37	57
HUDDLESTON, JG	1	3	2	18.12	73.34	0	0.11	59	201
YOSHIDA, T	1	3	2	5.05	27.46	0	0	8	72
TUNDO, P	1	2	2	20.02	30.42	0	0	64	154
KOHNO, K	1	1	1	12.87	12.87	0	0	62	62
YASUDA, H	1	1	1	7.34	7.34	0.04	0.04	159	159
SELVA, M	1	1	1	20.02	20.02	0	0	64	64
SHAIKH, AAG	1	1	1	12.04	12.04	0.03	0.03	65	65
SIVARAM, S	1	1	1	12.04	12.04	0.03	0.03	65	65
BORNER, A	1	1	1	0	0	0.05	0.05	17	17
SCHAFFNER, B	1	1	1	0	0	0.05	0.05	17	17
SCHAFFNER, F	1	1	1	0	0	0.05	0.05	17	17
VEREVKIN, SP	1	1	1	0	0	0.05	0.05	17	17

connected with recent and transient literature, while editorials were focused on presenting the advancements and milestones in GC – which corroborates the fact that almost 60% of intellectual authorities are related to the editorials. It should also be taken into account that editorials are structured around principles, not specialties or research themes, which may explain why none of the hubs or intellectual authorities of big specialty B - Ionic Liquids are related to the editorials, as well as account for the great participation of author of specialty E – GC Characterization.

5   Final Remarks

Table 14 presents a summary of the most important authors in each specialty. Authors in the research fronts from each specialty were assessed by the accumulated coverage of their output, generating the roll of 14 researchers that act as intellectual hubs of GC specialties. As the research front is about the most relevant and fresh research in certain time, these intellectual hubs are the major responsible for divulgating, sharing and organizing an specialty around an research object. The knowledge in the research fronts are built upon the previous selected knowledge, described in the intellectual base, from where authors with high significance functions as intellectual authorities to their specialties. Twenty-one researchers contribute with fundamental knowledge to sustain certain GC specialties. Surely, those names (either hubs or authorities) are not the only ones structuring the GC field, neither they are responsible for all the knowledge in this research branch. But they have special significance in funding and spreading the GC practices.

Table 14. GC hubs and intellectual authorities and their specialties.

Specialties	Hubs	Authorities
A – Solvents	Victorio Cadierno Javier Francos	Chao-Jun Li
B – Ionic Liquids	Robin D. Rogers Kenneth R. Seddon	Kenneth R. Seddon Robin D. Rogers Tom Welton Peter Wasserscheid
C – Biomass	David M. Alonso James A. Dumesic Atsushi Takagaki Michikazu Hara	James A. Dumesic George W. Huber Avelino Corma Sara Iborra
D – Catalysis	Rajender S. Varma Michikazu Hara	Rajender S. Varma André Loupy Vivek Polshettiwar
E – GC Characterization	Paul T. Anastas Roger A. Sheldon James H. Clark	Roger A. Sheldon James H. Clark David J. C. Constable Alan D. Curzons Barry M. Trost
F – CO_2 as substrate	Johannes Steinbauer Thomas Werner	Philip G. Jessop Walter Leitner Toshiyasu Sakakura Jun-Chul Choi

This research allows for a better understanding of the structure of GC and can assist researchers in searching for relevant information about topics of interest. Co-citation analysis was a suitable strategy for the analysis of groupings, and the use of its metrics showed to be relevant for determining hubs and intellectual authorities in the field. However, the lack of author keywords in some journal, such as Green Chemistry, proved to hurdle the information retrieval. Also, the changing of algorithm for Keywords Plus in

Web of Science makes it difficult to retrieve the same registers nowadays. A thoughtful standardization of databases and indexation is required and is a fact worth taking into account in future research. Comparisons with qualitative descriptions made in the field corroborate the coherence of the results of this research, and eventual discrepancies seem to reflect differences in the approach to information in the field: the analyses made by experts start from the Twelve Principles as prior categories for the organization of the GC, while this research draws specialties from the field's own citation patterns. This seems to be a more coherent strategy with the current GC target of thinking the Twelve Principles in a comprehensive, interdependent way, overcoming the incremental vision based on isolated principles [8].

References

1. ACS: History of Green Chemistry. https://www.acs.org/content/acs/en/greenchemistry/what-is-green-chemistry/history-of-green-chemistry.html. Accessed 28 Sep 2019
2. Anastas, P.T., Han, B., Leitner, W., Poliakoff, M.: "Happy silver anniversary": green chemistry at 25. Green Chem. **18**(1), 12–13 (2016). https://doi.org/10.1039/c5gc90067k
3. Poliakoff, M.: Green chemistry: science and politics of change. Science **297**(5582), 807–810 (2002). https://doi.org/10.1126/science.297.5582.807
4. US EPA, O.: Pollution Prevention Law and Policies. https://www.epa.gov/p2/pollution-prevention-law-and-policies. Accessed 28 May 2018
5. Anastas, P.T., Warner, J.C.: Green Chemistry: Theory and Practice. Oxford University Press, Oxford (1998)
6. Marcelino, L.V., Pinto, A.L., Marques, C.A.: Scientific specialties in green chemistry. Scientometrics (submitted)
7. Marcelino, L.V., Marques, C.A.: Trends in green chemistry research. In: Reunião Anual da Sociedade Brasileira de Química: programa e resumos, p. 875. SBQ, Joinville (2019)
8. Anastas, P.T., et al.: The Green ChemisTREE: 20 years after taking root with the 12 principles. Green Chem. **20**(9), 1929–1961 (2018). https://doi.org/10.1039/c8gc00482j
9. Clark, J.H., Sheldon, R.A., Raston, C., Poliakoff, M., Leitner, W.: 15 years of green chemistry. Green Chem. **16**(1), 18–23 (2014). https://doi.org/10.1039/c3gc90047a
10. Ivanković, A.: Review of 12 principles of green chemistry in practice. Int. J. Sustain. Green Energy **6**(3), 39 (2017). https://doi.org/10.11648/j.ijrse.20170603.12
11. Chen, C.: CiteSpace II: detecting and visualizing emerging trends and transient patterns in scientific literature. J. Am. Soc. Inf. Sci. Technol. **57**(3), 359–377 (2006). https://doi.org/10.1002/asi.20317
12. Chen, C.: Science mapping: a systematic review of the literature. J. Data Inf. Sci. **2**(2), 1–40 (2017). https://doi.org/10.1515/jdis-2017-0006
13. Kleinberg, J.M.: Hubs, authorities, and communities. ACM Comput. Surv. **31**(4), 1–5 (1999). https://doi.org/10.1145/345966.345982
14. RSC: "Happy silver anniversary": Green Chemistry at 25. Themed Collection. https://pubs.rsc.org/en/journals/articlecollectionlanding?sercode=gc&themeid=17fdd7fc-ff5b-46ca-ba3e-5702b2eb223. Accessed 24 Sep 2018
15. Noyons, E.C.M.: Bibliometric mapping as a science policy and research management tool (1999). https://openaccess.leidenuniv.nl/handle/1887/38308
16. Price, D.J.D.S.: Networks of scientific papers. Science **149**(3683), 510–515 (1965). https://doi.org/10.1126/science.149.3683.510
17. Anastas, P.T., Eghbali, N.: Green chemistry: principles and practice. Chem. Soc. Rev. **39**(1), 301–312 (2010). https://doi.org/10.1039/b918763b

18. Sheldon, R.A.: Green chemistry and resource efficiency: towards a green economy. Green Chem. **18**(11), 3180–3183 (2016). https://doi.org/10.1039/c6gc90040b
19. Delidovich, I., Palkovits, R.: Catalytic versus stoichiometric reagents as a key concept for green chemistry. Green Chem. **18**(3), 590–593 (2016). https://doi.org/10.1039/c5gc90070k
20. Li, C.J.: Reflection and perspective on green chemistry development for chemical synthesis— Daoist insights. Green Chem. **18**(7), 1836–1838 (2016). https://doi.org/10.1039/c6gc90029a
21. Llevot, A., Meier, M.A.R.: Renewability – a principle of utmost importance! Green Chem. **18**(18), 4800–4803 (2016). https://doi.org/10.1039/c6gc90087a
22. Quadrelli, E.A.: 25 years of energy and green chemistry: saving, storing, distributing and using energy responsibly. Green Chem. **18**(2), 328–330 (2016). https://doi.org/10.1039/c5g c90069g
23. Peters, M., von der Assen, N.: It is better to prevent waste than to treat or clean up waste after it is formed – or: what Benjamin Franklin has to do with "Green Chemistry". Green Chem. **18**(5), 1172–1174 (2016). https://doi.org/10.1039/c6gc90023b
24. Wakaki, T., Oisaki, K., Kanai, M.: Elementary and systemic views of the generation of toxic substances. Green Chem. **18**(13), 3681–3683 (2016). https://doi.org/10.1039/c6gc90058e
25. Sneddon, H.: Safety first. Green Chem. **18**(19), 5082–5085 (2016). https://doi.org/10.1039/ c6gc90086k
26. Jessop, P.G.: The use of auxiliary substances (e.g. solvents, separation agents) should be made unnecessary wherever possible and innocuous when used. Green Chem. **18**(9), 2577–2578 (2016). https://doi.org/10.1039/c6gc90039a

Characterization of Women's Scientific Participation in Brazil

Monique de Oliveira Santiago[(✉)], Thiago Magela Rodrigues Dias, and Felipe Affonso

Centro Federal de Educação Tecnológica de Minas Gerais (CEFET-MG),
Belo Horizonte, Brazil
moniqueosantiago@gmail.com, thiagomagela@gmail.com,
felipe-affonso@hotmail.com

Abstract. Several researchers have concentrated efforts to study the participation of women in scientific and technological careers, seeking to profile their trajectory and academic performance in science. In this context, this study aimed to analyze the participation of the group of doctor's degree who have curricula registered in the Lattes Platform and whose registered gender is female. After data collection, the stage of selection of curricula by gender criteria was performed, and data processing obtained a set of 149,850 registered curricula with female gender and maximum completed doctor's degree distributed in its various areas of scientific knowledge. The doctor's degree data were grouped regarding academic background, publications, productions, orientations, major areas of expertise and it was possible to analyze the evolution of scientific and technological production of the set in a temporal way. Studying the various aspects of gender difference in general and particularly in science and technology, as well as being relevant, can be a source of inspiration for government policies and programs that seek to promote change, encourage and value women's participation in science.

Keywords: Women in science · Lattes Platform · Scientific and technological production

1 Introduction

Scientific production has grown significantly in recent years, using the Web as one of the means that facilitates access and dissemination of its content. The data regarding scientific production has provided studies that seek to understand how science has evolved and how scientific collaboration occurs. Among the various studies applied to this set, one that has gained prominence is gender studies. Gender can be defined as the connection of two propositions: "[...] gender is a constitutive element of social relationships based on perceived differences

© ICST Institute for Computer Sciences, Social Informatics and Telecommunications Engineering 2020
Published by Springer Nature Switzerland AG 2020. All Rights Reserved
R. Mugnaini (Ed.): DIONE 2020, LNICST 319, pp. 210–221, 2020.
https://doi.org/10.1007/978-3-030-50072-6_16

between the sexes, and gender is a primary way of signifying relationships of power" [9, p. 1067], therefore, corresponds an interdisciplinary field that has as its theme the identity and representation of men and women in society. This field includes the subfield Women's Study, which covers, among various dimensions of interest, women and their various relationships with science [5].

Because it is an interdisciplinary and comprehensive theme, studies that focus on women and their varied relationships with science have several approaches that seek to outline a career trajectory and female performance. Despite progress in women's participation in various segments in their academic and scientific careers, there is still a gender gap in science around the world that needs to be better understood. There is clear evidence of gender inequality in Brazilian Science, which probably has deep institutional and cultural roots [8]. Thus, conducting a study to investigate the scientific participation of women in Brazil is a step to understand the current scenario, and through this understanding contributes to the adoption of measures that promote gender equality. There may be differences in the performance of men and women in Brazilian science, but these differences are related to their presence in the field and the clipping that is given to the analyzes [5]. Reflecting on and searching for new ways to investigate women's productivity in science is a step towards minimizing existing career inequalities.

One of the biggest difficulties in analyzing a country's scientific production may be related to the acquisition of data, which is usually present in several repositories. However, this process can be facilitated by using the Lattes Platform curriculum data repository. The Lattes Platform is considered an important set of Brazilian scientific data, in which it provides high-quality information and enables researching data from individuals who are registered there, such as academic background and scientific production, among others [4]. This data set integrates into a single system the databases of Curriculum, Research Groups and Educational Institutions of the country and has curriculum information of the academic trajectory of the entire Brazilian scientific community.

Studying large data repositories becomes a complex task because the amount of data to be analyzed and the characteristics of each repository are unique and most have no definite pattern. In this scenario, bibliometrics emerges that seeks to quantify the processes of written communication, using methods for statistical analysis of the production and dissemination of knowledge applied to scientific data sources [1]. Therefore, bibliometric analyses are used as indicators of scientific production to provide indicators for national planning for the evolution of scientific research.

In this context, this paper aims to analyze the scientific and technological women's participation, investigating how their research has been carried out and how their academic career has evolved over the years from bibliometric analyses performed on curriculum data available on the Lattes Platform. This study, besides presenting an overview of women who have been conducting research, aims to contribute to the generation of national scientific indicators and the management of information in the scientific and technological areas.

One of the factors that motivate the study or understanding of women's scientific participation is the opportunity to understand the current scenario, and through this way to implement measures to promote reproductions between genders, minimizing possible career inequalities. Since, by recognizing that " [...] a science is gender-neutral, revealing the values and social characteristics attributed to women are devalued in the production of knowledge, and that gender inequalities pervade the field. scientific [...] " [10, p. 464, our translation][1], admits a need to break down as barriers to inequality and thus ensures research excellence by retaining its best researchers and innovators [8] and thus enables " [...] to create a science more human, free from the transformations caused by centuries of exclusion of half of humanity as women " [5, p. 150, our translation][2].

2   Development

All curriculum information for the Lattes Platform is included by the researcher himself and is freely available on the internet. This broad data set contains the entire record of the researcher's professional, academic and scientific career. Due to its wealth of information and as this data source has not been widely analyzed, it is therefore justified to choose the Lattes Platform as a data source to measure and evaluate women's national scientific performance.

The process of extracting and selecting of curriculums data from the Lattes Platform was performed through the LattesDataXplorer framework [3], which has a set of techniques and methods responsible for collecting, selecting, processing and analyzing the data. Thus, the framework collection module was responsible for collecting all curriculum in March 2019, surpassing 6,126,000 records.

LattesDataXplorer was specifically used for the collection and selection of the Lattes Platform of curriculums data, which obtained the Doctor's Degree Curriculums Repository in XML. After this process, it was necessary to select data from this repository and thus obtain the repository of all women's curriculums with completed academic/doctor's degree. Finally, the data were processed and visualized to analyze the scientific participation of women (Fig. 1).

To carry out a detailed analysis of women's national scientific participation, it was decided to limit the data through the level of academic education/degree, reducing the set for individuals who have completed the doctor's degree level. Although this set is not the most significant among the levels of education, they account for 74.51% of papers published in journal article and 64.67% published in conference paper, besides have generally updated date of their curriculums recently and notably are responsible for the highest level of training, namely,

---

[1] [...] a ciência é neutra com relação às questões de gênero, revelando que os valores e as características socialmente atribuídos às mulheres são desvalorizados na produção do conhecimento, e que desigualdades de gênero perpassam o campo científico [...].

[2] [...] criar uma ciência mais humana, livre das transformações causadas pelos séculos de exclusão de mais da metade da humanidade, as mulheres.

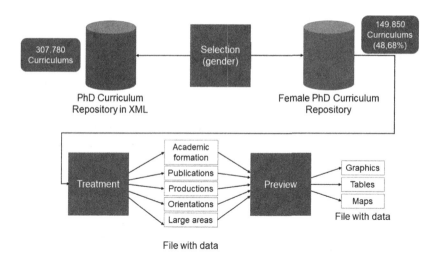

Fig. 1. Data selection, treatment and preview process.

masters and doctor's degree [3]. Therefore, after the acquisition of all curriculums registered in the Lattes Platform, the LattesDataXplorer selection module was used to select, among them, the curriculums that have completed the doctor's degree, thus totaling a set of 307,780 (5,02%) curriculums.

After the curricular data collection and doctor's degree selection stages, the selection by gender criteria was performed, as some fields collected by the Lattes Platform are not displayed in the public consultation, such as CPF, gender, color or race, birth data, identity, passport, membership, home address, and more. Thus, we selected the female-owned curriculum from the Phd Curriculum Repository in Xml yielded 149,850 female curricula. This total corresponds to 2.45% of all Lattes Platform data.

The XML files extracted from the Lattes Platform have a well-defined structure, with structured information and data diversity, such as level of academic education/degree, large areas of expertise, research and extension projects, bibliographic and technical productions such as articles published in annals congress and papers in journal, presentation of papers, participation in newsstands, events, orientations, among others. As each curriculum has a specific amount of this information, treatment was performed to improve understanding of the data. Therefore, the preprocessing step was performed by accessing each XML curriculum, obtaining that curriculum-specific information, and storing it in a collection of structured files. In the end, these data were grouped in terms of academic background, publications, productions, orientations, major areas, among others. After grouping the data, a characterization was performed to facilitate the preview and analysis of the data. In this step, for each collection of structured files were generated charts, tables, maps, among others.

3   Results

3.1   General Characterization

Data collected from the Lattes Platform using the LattesDataXplorer framework in March 2019 totaled more than 6,126,000 records. From this total, the records with the level of academic doctor's degree completed were selected, totaling 307,780 (5.02%) curriculums from the various areas of scientific knowledge. These same data were selected by gender criteria, in which 149,850 (48.68%) correspond to curriculums of doctor's degree female and 157,942 (51.31%) to curriculums of doctor's degree male.

This data shows significant growth regarding the data provided by the Lattes Panel [7]. The last extraction performed by Lattes Panel took place on November 30, 2016, being the same made available for public consultation, which provided data regarding the panel Distribution by gender, age and area corresponding to 63,853 (47.50%) doctor's degree curriculums female and 70,567 (52.49%) doctor's degree curriculums male. Thus, by comparing these data provided by the Lattes Panel with the current data collected from the Lattes Platform, we can see the growth of 128.97% of researchers who completed their doctor's degree in more than two years. Another relevant aspect refers to the percentage of doctor's degree training, in which females increased by 1.18% and males decreased by −1.38%. Even with this considerable percentage, it was not enough for female doctor's degree to outnumber male doctor's degree.

With the general characterization of the data, it was possible to visualize the representativeness of the selected set within the total set of records. As the focus of this work is to characterize the Brazilian female scientific participation, the next analyzes will be used only the data set corresponding to the doctor's degree curriculums female (149,850 curriculums), which corresponds to 2.45% of the total of all curriculus registered in the Lattes Platform.

3.2   Women's Orientations Profile

Relevant information for analysis corresponds to the completed and ongoing orientations of women, as they are the doctor's degree responsible for the education of students at different levels of education. The completed orientations correspond to all the orientations made by the doctors since the beginning of their career and which are already finalized, while the orientations in progress are those that are under development and not yet finalized. Thus, by quantifying these data we have a total of 3,577,801 completed orientations and 341,607 ongoing orientations (Fig. 2).

An important aspect in the graphs above refers to the levels of postgraduate education: master's degree, doctor's degree, and postdoctorate. At these three levels, the sum of the percentages for the completed orientations graph is 17%, while for the ongoing orientations graph this sum is 47%. One hypothesis for this percentage difference concerns the fact that, as doctor's degree are responsible for student education in major postgraduate programs *stricto sensu* in Brazil,

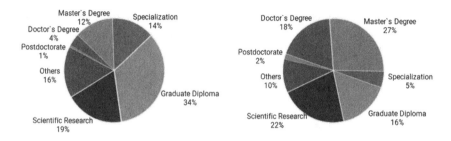

Fig. 2. Completed orientations (left) and ongoing orientations (right).

they tend to mentor more undergraduate students and fewer postgraduate students at the beginning of their careers and over the years after completing their doctorate the number of undergraduate counseling decreases and the postgraduate counseling increases. However, as all orientations are considered throughout their history, the completed orientations have the lowest level, as undergraduate degrees being more representative.

3.3 Representativeness of Women's Bibliographic and Technical Production

The doctor's degree are responsible for most of the works registered in the Lattes Platform [3], so a survey of the main bibliographical, technical and other artistic/cultural productions of women was carried out. Thus, verifying the scientific and technological production of the data set of female doctor's degree, we arrive at the quantitative, presented in (Table 1), corresponding to the main productions.

Table 1. Quantitative data by bibliographical and technical productions of womans doctor's degree.

Production type	Production
Conference paper	5.092.804
Journal article	2.274.378
Other technical productions	1.375.610
Technical works	824.913
Book chapters	593.484
Texts in newspapers and magazines	314.212
Other bibliographic productions	313.423
Books	200.515

The previous table shows that the number of papers published in a conference paper and journal article is higher than the other types of productions, both of which are the most common for publication and dissemination of content. From this data and taking into account the productivity of the doctor's degree per year, a quantitative was performed (Fig. 3) with the most relevant types of productions.

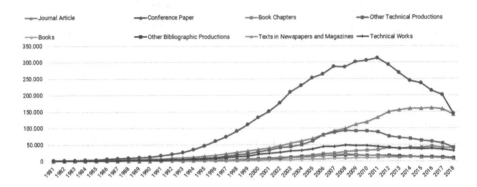

Fig. 3. Quantitative of scientific productions per year.

The first twenty years (1961 to 1981) scientific production remained constant for all types of productions. Possibly, a hypothesis to explain this fact may be related to the launch and standardization of the Lattes Platform curriculum that took place in August 1999, in which data referring to publications prior to this period may not have been disclosed by the doctor's degree. Another hypothesis that may influence this fact is related to the number of doctor's degree, who were few at the time.

Since the early 1981s, there has been an increase in all types of scientific and technological production. The productions that emerged from the others correspond to journal article and conference paper. Regarding journal article, it presented a significant growth until 2013, remaining constant after this year and falling in 2017. The production of conference paper had a considerable increase with the apex in 2011 and a significant decrease after that year. This same behavior is presented in the study of [3], for all doctor's degree with curriculums registered in the Lattes Platform. This steep decline in conference paper was so significant that from its peak until 2018, it fell by 53.62%, reaching the end of last year with a total value of articles close to the total of journal article.

Different hypotheses may be related to the steep slope of the conference paper from 2011. One of them refers to the classification of scientific production used by CAPES, if we consider that the production evaluation system influences the actions of individuals. This fact is driven by the non-consideration of articles in conference paper in postgraduate program evaluations. Thus, journal article becomes more interesting for publication, as they influence the concepts of the programs in which the doctor's degree participate and their efforts for this type of publication.

3.4   Distribution of Women in Their Major Areas

The information that describes the professional ties can be registered in the Lattes Platform curricula through the "Practice/Practice Areas" menu, allowing to choose and inform numerous knowledge areas and subareas. The large area can be defined as a "clustering of diverse areas of knowledge by virtue of the affinity of their objects, cognitive methods and instrumental resources reflecting specific sociopolitical contexts" [2]. The Lattes Platform has nine options for large areas that aggregate their respective areas, following the CNPq classification[3].

To perform this analysis, the first option filled out by the female doctor's degree was selected considering that this is the most relevant area among the others. Thus, using the two most significant means of production to group the doctor's degree curriculum by major practice areas (Fig. 4), as expected, resulted in the most significant major human-related fields and the least significant corresponding to engineering. These data showing that men predominate in exact careers and women predominate in biological areas and health [6].

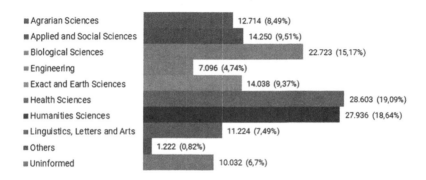

Fig. 4. Major areas of expertise.

A significant number of curriculums do not have information related to their area of expertise (10,032 curriculums). This can be explained by the fact that, as it is not mandatory information, when entering a new academic formation there is no concern to fill in the data of this field, or even there may be forgetfulness on the part of doctors. Another aspect that can be taken into account refers to the difficulty factor in associating the course with the area of expertise, especially for interdisciplinary courses and recent postgraduate courses, which results in not informing data for this field.

As journal article and conference paper are the two most significant means of production, they were used to verify scientific production by a large area of expertise (Fig. 5).

---

[3] CNPq Knowledge Area Table: http://lattes.cnpq.br/documents/11871/24930/Tabel adeAreasdoConhecimento.pdf/d192ff6b-3e0a-4074-a74d-c280521bd5f7.

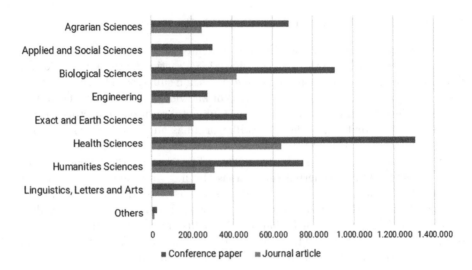

Fig. 5. Journal article and conference paper by area of activity.

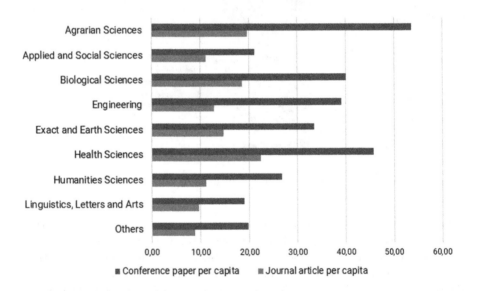

Fig. 6. Journal article and conference paper per capita by area of activity.

By dividing the total number of articles by the total number of individuals according to the corresponding large area of activity, we have a number of productions per capita (Fig. 6). This analysis becomes more interesting because it is possible to verify the production without the influence of a large number of authors from some areas.

Comparing the two previous figures, we can see a considerable change of positions between some areas of expertise in the production of journal article and conference paper. For the articles in conference paper, the first area that stands out about the others is the Agrarian Sciences, in which Fig. 5 presents itself as the fourth-largest producer of articles and Fig. 6 appears first. Health Sciences and Biological Sciences, which were first and second in Fig. 5, remain quite significant in Fig. 6, second and third place respectively. The Engineering area had a surprising and unexpected change from the second to last (Fig. 5) to fourth place (Fig. 6). The Humanities Sciences that was in third, moved to sixth place.

For journal article, the change of positions appears less significant than for articles in conference paper. Featured for Agrarian Sciences from a fourth-place (Fig. 5) appears second in Fig. 6., and the Engineering area had a surprising and unexpected change from the last (Fig. 5) to the fifth (Fig. 6).

These data reveal to us the importance of the scientific production of the doctor's degree in several major areas. It was possible to identify the current scientific production scenario and to verify that even the areas with less female participation, such as engineering, per capita production presents the opposite scenario.

3.5 Evolution of the Academic Formation of Female Doctor's Degree

All academic history can be informed in the Lattes Platform curricula, allowing to verify the academic trajectory of the doctor's degree. For doctor's degree, master's degree and professional master's degree, one can inform the major areas of knowledge and choose from nine options by selecting the areas and then the subareas that relate to the postgraduate course.

As it is possible to fill in more than one area of knowledge for each postgraduate informed, for this analysis was selected only the first area filled by the female doctor's degree, considering to be the most relevant among the areas registered. Thus, it is possible to verify the evolution of the major areas for the doctorate academic formation (Fig. 7).

As some fields in the Lattes Platform curriculum are not required, data for the starting year, ending year, or large area of knowledge that did not contain information was characterized as Uninformed and these records were not used for analysis. Noting that for the doctorate 45,510 (30.37%) records were categorized as Not informed.

The 71 years of history regarding the formation of doctors in the Lattes Platform were divided into four periods. In these periods, the major areas that formed the most doctor's degree per period alternated between Exact and Earth Sciences, Humanities Sciences, Biological Sciences, and Health Sciences.

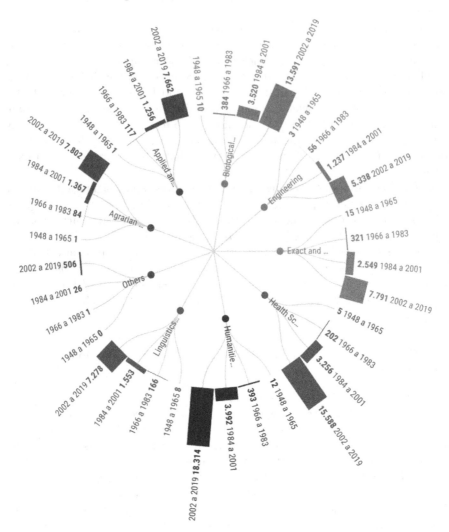

Fig. 7. Evolution of the major areas for the doctorate academic formation.

4   Final Considerations

Studies that focus on the various aspects of gender difference in work in general, and particularly in science and technology, are relevant and can be a source of inspiration for government policies and programs that promote change and lead to equal participation for women and men. As this study presents the full potential offered by the Lattes Platform for the understanding of women's scientific participation in Brazil, all the results presented aimed to present a view of female participation in a temporal way on the data set. As future work is expected to perform an analysis of the scientific collaboration of doctor's degree, CNPq research productivity scholarship holders and the scientific production of doctor's degree working in postgraduate programs.

Supported: The authors thank CAPES for their research assistance.

References

1. Araújo, C.A.A.: Bibliometria: evolução histórica e questões atuais. Em Questão **12**(1), 11–32 (2006)
2. CAPES. Tabela de Áreas de Conhecimento/Avaliação, Brasil (2019). https://www. capes.gov.br/avaliacao/sobre-a-avaliacao/mestrado-e-doutorado-o-que-sao/44-cont eudo-estatico/index.php?option=comcontent&view=article&id=6831:tabela-de-ar eas-de-conhecimentoavaliacao&catid=91:avaliacao-capes&Itemid=759. Acesso 04 dez 2019
3. Dias, T.M.R.: Um Estudo da Produção Científica Brasileira a partir de Dados da Plataforma Lattes. Centro Federal de Educação Tecnológica de Minas Gerais, Doutorado em Modelagem Matemática e Computacional (2016)
4. Lane, J.: Let's make science metrics more scientific. Nature **464**(7288), 488 (2010)
5. Leta, J.: Mulheres na Ciência Brasileira: desempenho inferior? Revista Feminismos **3**(2), 139–152 (2014)
6. Olinto, G.: A inclusão das mulheres nas carreiras de ciência e tecnologia no Brasil. Inclusão Soc. **1**(5), 68–77 (2011)
7. Painel Lattes: Distribuição por Sexo, Faixa Etária e Grande Área de Atuação. http://estatico.cnpq.br/painelLattes/sexofaixaetaria/
8. Valentova, J.V., Otta, E., Silva, M.L., McElligott, A.G.: Underrepresentation of women in the senior levels of Brazilian science. PeerJ **5**(1), 1–20 (2017)
9. Scott, J.W.: Gender: a useful category of historical analysis. Am. Hist. Rev. **5**(91), 1053–1075 (1995)
10. Silva, F.F., Ribeiro, P.R.C.: Trajetórias de mulheres na ciência: "ser cientista" e "ser mulher". Ciência Educação **2**(20), 449–466 (2014)

An Altmetric Alternative for Measuring the Impact of University Institutional Repositories' Grey Literature

Miguel Valles$^{(\boxtimes)}$ ⓘ, Richard Injante ⓘ, Edwin Hernández ⓘ, Juan Riascos ⓘ, Marco Galvez ⓘ, and Juan Velasco ⓘ

Universidad Nacional de San Martín, Tarapoto, Peru
mavalles@unsm.edu.pe

Abstract. To get the license, universities must meet a set of standards to guarantee the education of their students. This involves installation and maintenance of an institutional repository to publish their grey literature. This research aims to create an alternative altmetric tool to measure the impact of such literature. To get it done, we worked the institutional repository customization and optimization, we created procedures to retrieve records, transform and load altmetric data in a new database and we built a dashboard-type tool using the altmetric data of the institutional repository grey literature. Now, there is an alternative to measure the altmetric indices from different dimensions and analysis perspectives and it makes possible to take the necessary actions to apply strategies that aim to increase the repository visibility. The tool implementation in the study case allowed the monitoring and control of the institutional repository altmetric indices.

Keywords: Altmetric · Institutional repositories · Grey literature · Bibliometric · Dashboard

1 Introduction

Nowadays, Latin America and world universities are required a set of demands to fulfill the basic conditions of educational quality that has forced them to guarantee better conditions for a comprehensive education of professionals graduating from their schools. The trends is to look for a less heterogeneous, less unequal university scenario towards the future, but above all it is sought that they assume the role that allows building a society with better living conditions for its population, a democracy and economies strengthened as a result of increase science, education and culture [1]. Peru is no exception. In that sense, the new university law No. 30220 [2] has had a special impact on process of professionals education by demanding that universities comply with 55 minimum standards, which form a protection mechanism for students, their families and society as a whole [3, 4].

Part of the requirements implies research results publication in open access through institutional repositories to be used free of charge as a background for research or to

R. Mugnaini (Ed.): DIONE 2020, LNICST 319, pp. 222–234, 2020.
https://doi.org/10.1007/978-3-030-50072-6_17

strengthen the bibliographic review of a topic to be investigated. Once the research results are available, the design of strategies that allow to increase its visibility and impact factor remains, incorporating as part of repository functions the sharing of resources in social networks as a marketing strategy to achieve a greater audience that the academic environment itself.

In that sense, Peruvian government has been making efforts along with universities to give licensing, demanding updated and accessible repositories [4], so that the revision of their content serves to formulate research projects leading to academic and/or professional degrees. This is done through the National Register of Research Works (RENATI from Spanish initials), the Free Access to Scientific Information for Innovation (ALICIA from Spanish initials) project, with Law No. 30035 endorsement that regulates the Open Access National Repository of Science, Technology and Innovation [5] and its corresponding regulations [6].

Therefore, Peruvian universities have been working to setup the infrastructure, specifications, as well as on the identification of requirements and processes training for the human resource responsible to guarantee that thesis and every material published in their repositories are available. Peruvian universities uses DSPACE to support these processes. The Standard 37, Condition V of the Licensing Model, requires the institutional repository maintenance where research documents like thesis, research reports and scientific publications, among others should be registered and stored. Ferreras-Fernandez et al. [7] state that these open access repositories are valid channels for grey literature publication. Grey literature are publications that do not go through a strict quality validation process such as peer revision or have a limited circulation. This literature is usually recommended as bibliographic revision source in research documents, but recommendation is to avoid it, nevertheless, for A. Paez [8] grey literature can make great contributions to a systematic revision.

Measurement of repository performance indices and measurement of quality and quantity bibliometric indices of scientific production used to take actions to increase these indicators are complicated to monitor without a proper knowledge of techniques and tools available for this purpose. This also involves design of strategies that guarantee the use of the available material, visibility increase and impact factor of publications using social networks as a marketing and advertising strategy. Thelwall et al. [9] state that references in social networks have become a valuable marketing tool for publishers who try to advertise their articles. For S. Santana [10] current research processes are increasingly collaborative and communication between academics and researchers is developed through social interchange spaces, and this accelerates the diffusion of scientific knowledge, so the job of diffusion in social networks is necessary. In addition, according to the altmetric manifesto [11] since altmetrics are in themselves diverse, they are excellent at measuring the impact of the diverse academic ecosystem.

According to Priem et All [12], altmetrics is the study and use of academic impact measures based on activity in online tools and environments. The term describe the metrics themselves—one could propose in plural a "set of new altmetrics." Altmetrics is in most cases a subset of both scientometrics and webometrics; it is a subset of the latter in that it focuses more narrowly on scholarly influence as measured in online tools and environments, rather than on the Web more generally. The sources used for altmetrics are heterogeneous and include mentions and citations in blogs, Wikipedia, Twitter, Facebook, Mendeley or reader counts on social reference managers, bookmarking platforms and bibliographic references software. In that sense I. Aguillo [13] mentions that altmetrics is being explored as a potential tool for measuring research impact beyond the scientific communities, the so-called societal impact.

Why use altmetrics then? According to L. Palmer [14], because it provides new services and demonstrates content evaluation and assessment, integrates social network data, shows social importance of our content, adds context and meaning to resources downloads, plus the badges and API are free. L. Palmer [14] also mentions the concern about the future costs of aggregate exploitation services of the altmetric data and its complexity of use, justifying once more the need to create a dashboard based monitoring tool that makes use of the Altmetrics API functions.

The possibility for using social networks to divulge the scientific production results, specifically for repository's grey literature, has already been implemented in Peru (as shown in L. Elespuru & L. Huaroto [15] study). Thanks to configuring features for resource sharing on social networks and incorporating free altmetrics donuts provided by altmetrics.com [16] they have increased their visits, reading, quotations and impact factor according to E. Adie & W. Roe [17]. It has also allowed them complementing existing usage statistics to help plan repository development and the allocation of resources to improve marketing strategies for its content [18].

This implies the need to measure, control and improve the altmetric and bibliometric indices of institutional repository's grey literature. Many repository managers create in-house–generated spreadsheets and monthly statistics information that cannot be tracked easily or efficiently by repository software [19]. A way to achieve this is through subscription to services of existent alternative metrics, for example: Altmetrics, PlumX, Sciencecard, SciVerse Scopus API and others [9, 19, 20]. However, due to its high costs, complexity of initial setup work and data loading, it has become a distant possibility for universities. They want not just to integrate use of social networks as a marketing strategy to advertise their resources but they want use the information of relevant discussions of their content, analyze and compare the scope that dependencies have achieved within the organization or know the audience demographic data that only paid subscription services provide.

Another way, without paid subscription, is using an application-programming interface (API) which includes routines, data structures, objects classes, variables and response codes that subscription services provide. It is free, it simplifies the process and allow access to Altmetrics.com database for free [9] to those who wish to understand the patterns of academic communication in the web [21], but their use conditions force attribution to source from which data has been mined.

During literature review, we found proposals such as Zervas et All's [22] who converted the Cyprus University of Technology's repository into a Current Research Information System (CRIS). They mention that altmetrics is already implemented in the CRIS and their most important result is that now they have an institutional repository that can collect, manage, preserve and disseminate all information about their university's research and its performance. Also Konkiel & Scherer's [18] who contextualized the importance of use Altmetrics for measurement of valuable indicators of interest to complement statistics and traditional use, however they also mentioned the services cost (as equal as L. Palmer [14]) to obtain aggregate indicators of Altmetric information that generates the use of its resources in social networks. And finally I. Aguillo's [13] who mentions that repositories play a key role in success of open access publishing. However, he indicates that making them available in this format is not enough to achieve visibility and impact, so it is important to use both public and academic social networks to achieve larger audiences and increase not only academic but also social impact.

However, we do not identified proposals to measure grey literature impact of university institutional repositories in social networks on the way that we can analyze and compare metrics graphically grouped by faculty, dependency, social network, demographics and any other indicator that has been achieved internally in the university without having to pay for the service. Therefore, this work proposes an alternative to integrate altmetric functions within repository and building a tool to monitor the altmetric indices using the API for developers provided by altmetric.com.

As a study case, we implemented the proposal with Universidad Nacional de San Martin repository, achieving improvements on the impact and visibility of published research.

2 Materials and Methods

This work provides a dashboard based monitoring tool that institutional repository managers can use to monitor mentions in social networks of available resources. For this, we customized the institutional repository website optimizing it and incorporating it altmetric functions; then we mined the repository records that serve as income to perform requests to the altmetrics API, resulting in a dataset from those that have been shared. Finally, we built the dashboard interface.

We used data and website of Universidad Nacional de San Martin – Tarapoto institutional repository, where we incorporated function to share the available resources (mainly grey literature) in social networks and we deploy the dashboard based monitoring tool in the following URL: http://altmetrics.unsm.edu.pe. Up to September 2019, there were 2870 bibliographic resources available. Below, we described and outlined the stages that we have followed to achieve the objectives of the work.

The steps necessary to build, configure and use the monitoring tool are parameterizable and standard, thus ensuring that any institution that uses DSPACE as its repository manager can use it.

2.1 Customizing the Institutional Repository

To share the resources available at repository and to be able to track them by alternative metrics services, it is necessary to perform some setups and including new functions that we detailed:

1. Institutional repository interfaces redesign to improve its downloading performance and usability.

2. Creating and setting up unique identification service for handle resources [18].
3. Incorporating complements to share resources in social networks using the thisadd.com service.
4. Integrating alternative metrics of altmetrics.com in the institutional repository.
5. Incorporating Google Analytics to monitor visits traffic to the institutional repository website.
6. Implementing optimization techniques to improve performance and positioning of repository website in search engines using Google Webmasters.
7. Creating and setting up services to update resources in Google Scholar.

2.2 Automated Mining, Transformation and Data Upload Procedures

This work is based on J. Ortega [20], Robinson-Garcia et all [23] and E. Adie & W. Roe [17]'s data mining model, who mine an initial set of articles from public pages that then were searched in altmetric.com, using its API [24] as data provider interface for scientometric developers and researchers [21, 25]. Currently, the version available according to Altmetric.com [24] is version 1.

We designed a database model in PostgreSQL that guarantees the structured storage of data that where pivoted and integrated from the DSPACE data model with the organizational structure created, as well as the results of Altmetrics API (from where it was possible to obtain records of the resources that have been shared in social networks). We show it below (Fig. 1).

We designed a standard, automated and parameterizable flow processes as follows (Fig. 2):

A query is created to mine every register of the repository's bibliographic resources which has a unique identification valid code; the university uses handle [26].

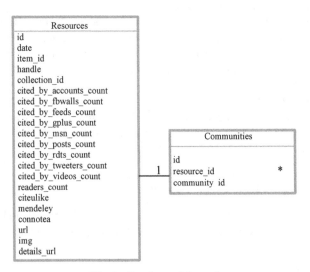

Fig. 1. Database of the tool

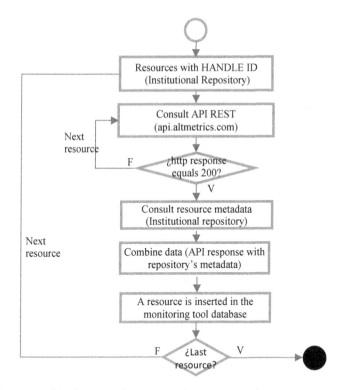

Fig. 2. Flow of standard and parameterizable processes

With this data an array is uploaded to perform timed requests to the Altmetrics API REST, then using the http methods available from the API REST a response is expected, which has one of the following codes [24] (Table 1):

Table 1. Requests response code.

Status code	Description
200	Success. The body of the response should contain the data you requested
403	You are not authorized for this call. Some calls and query types can only be made by holders of an API key and/or a license for the Full Access version of the API
404	Altmetric does not have any details for the article or set of articles you requested
429	You are being rate limited. If you have not already then apply for an API key
502	The Altmetric Details Page API version you are using is currently down for maintenance

If we found the resource in the altmetrics database, the API retrieves a metadata set in JSON format; then we combined it with the data obtained from a relational query to the institutional repository database to obtain detailed information and the resource hierarchical structure.

This combined data is stored in the database of dashboard based monitoring tool that we created.

2.3 Building the Monitoring Tool and Mining for Altmetric Data

Components of Monitoring Tool

The tool has three components: Data access, business logic and presentation.

Data access component contains normalized data of repository database and the ones mined from the API REST. Business logic component of the tool is an algorithm that implements flow of standard and parameterizable processes mentioned in point 2.3. For presentation component, we used the code editor Visual Studio Code that is compatible with the technologies: PHP, CSS, HTML5, Bootstrap, AngularJS and Laravel applying the Model View Controller (MVC) software design pattern to integrate all these technologies. We designed the interface very similar to https://www.altmetric.com/exp lore.

3 Results and Discussion

3.1 Customization of the Institutional Repository

First, the project requires repository customization; it includes user interfaces redesign, analytical functions incorporation with Google Analytics and repository website optimization with Google Webmasters. Now, repositories managers have a set of tools to monitor analytical data where there are different segments, metrics and analysis dimensions available [27].

The following graph shows analytical indices evolution of institutional repository website:

After customization of institutional repository website, we can see in Fig. 3 that visits increased by an average of 65% compared to previous period. With a minimum increase of 16% and a maximum of 156%. This indicates that interface redesign, social networks inclusion and site optimization suggested by Google Webmaster improved the repository visibility.

To determine statistically that the interface redesign, social networks inclusion and site optimization suggested by Google Webmaster improve the visibility of the repository we have:

Null hypothesis. H_0: $\mu_x \leq 0$ (If initial mean difference of repository interface redesign, social networks inclusion and site optimization suggested by Google Webmaster is less than or equal to zero, then the repository visibility is not improved).

Alternate hypothesis H_1: $\mu_x > 0$ (If final mean difference of repository interface redesign, social networks inclusion and site optimization suggested by Google Webmaster is greater than zero, and then the repository visibility improves) (Table 2).

When contrasting the hypothesis at 5% significance level, we have 138.5 as difference result, with 72.931 as standard deviation and with 95% confidence interval that the

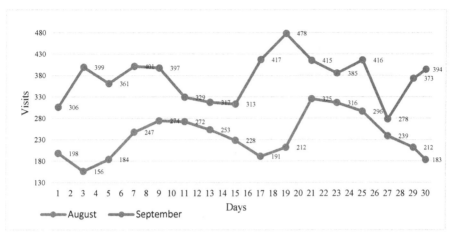

Fig. 3. Evolution of institutional repository website visits. August – September 2019 period.

Table 2. Paired samples test (SPSS.v.24)

	Paired differences					t	Df	Sig.
	Mean	Standard deviation	Mean standard error	95% TI				
				Lower	Higher			
Con_Interfaz-Convencional	138,5	72,931	18,233	99,638	177,362	7,596	15	*p = 0,000*

average difference is between [99.638; 177.632]. We conclude then that difference is significant with p = 0.000; that is to say the interface redesign, social networks inclusion and optimization of the site suggested by Google Webmaster significantly improves repository visibility.

3.2 Building of Automated Mining, Transformation and Uploading of Data Procedures

To use dashboard based monitoring tool, the system automatically perform recuperation process weekly on Sundays at 00:00. However, there is also the possibility to performing the procedure on demand. We can see the retrieved data in the following Table 3:

Table 3. Summary of retrieved data from altmetrics.

Date	Records in Dspace	Shared records	Times shared	Facebook	Twitter
08/12/2019	2,750	45	57	28	29
08/31/2019	2,767	51	64	34	30
09/17/2019	2,790	51	65	34	31
09/23/2019	2,801	51	65	34	31
10/05/2019	2,840	52	66	35	31
10/07/2019	2,857	57	71	40	31

3.3 Construction of the Altmetric Data Monitoring and Exploitation Tool

With this data, we obtained different kinds of queries in a graphic way. The most important ones are the following (Tables 4, 5, 6 and 7):

Table 4. Mentions in social networks summarized by university

Description	Mentions
At university level	71

Table 5. Mentions in social networks summarized by type of research

Description	Mentions	Percentage
Research and development	7	9.9%
Literary production	1	1.4%
Thesis	63	88.7%
Total	71	100.0%

Table 6. Mentions in social networks summarized by type of thesis

Description	Mentions	Percentage
Doctorate	2	3.2%
Master	2	3.2%
Undergraduate	59	93.7%
Total	63	63.0%

Finally, the monitoring tool interface, when used from a mobile device (Fig. 4).

Table 7. Mentions in social networks of undergraduate theses summarized by faculty

Description	Mentions	Percentage
Faculty of Agricultural Sciences	14	23,7%
Faculty of Health Sciences	2	3,4%
Faculty of Economic Sciences	6	10,2%
Faculty of Law and Political Sciences	0	0,0%
Faculty of Ecology	2	3,4%
Faculty of Education and Humanities	3	5,1%
Faculty of Agroindustrial Engineering	2	3,4%
Faculty of Civil Engineering and Architecture	6	10,2%
Faculty of Systems Engineering and IT	23	39,0%
Faculty of Human Medicine	1	1,7%
Total	59	100,0%

Fig. 4. Interface of monitoring tool "Observatory of the social networks"

4   Conclusions

We built a dashboard based monitoring tool were you may observe different indices of alternative metrics graphically, thanks to the mentions that repository resources get in social networks because marketing strategies, advertising and new academic uses given by the researchers.

We customized and optimized institutional repository website incorporating altmetric and analytical functions, allowing higher visibility increasing visits by 58% obtaining higher mentions in social networks, mainly Facebook and Twitter.

The tool implementation of study case allowed us monitoring and control institutional repository altmetric indices.

An important aspect of this alternative is the low cost for implementation and customization compared to subscription to existent alternative metrics services that has high costs and some complexity degree for customization.

Thanks to features, functional and non-functional requirements implemented, as well as the standardized flow of proposed processes, this tool may be implemented at any university institutional repository and thus obtain altmetric measurements of its grey literature.

Acknowledgements. To Universidad Nacional de San Martin – Tarapoto – Peru, for financing the research project called "Observatory of social networks and alternative metrics and their relationship with different bibliometric indices of institutional repository of UNSM-T" from which the results of this article arise.

References

1. Gazzola, A., Didriksson, A. (eds.): Trends in Higher Education in Latin America and the Caribbean. UNESCO. IESALC-UNE, Caracas (2008). 16–17 p.
2. MINEDU. Ley Universitaria 30220, p. 35. MINEDU, Lima (2015). http://www.minedu.gob.pe/reforma-universitaria/pdf/ley_universitaria.pdf. Accessed 27 Mar 27 2019
3. Sunedu. Condiciones básicas de calidad (2019). https://www.sunedu.gob.pe/8-condiciones-basicas-de-calidad/. Accessed 4 Oct 2019
4. Sunedu. El Modelo de Licenciamiento y su Implementación en el Sistema Universitario Peruano. Primera. Minedu, editor. MINISTERIO DE EDUCACION. SUNEDU, Lima, p. 37 (2015). http://repositorio.minedu.gob.pe/handle/123456789/4565
5. Ley N° 30035. Ley que regula el repositorio nacioanl digital de ciencia, tecnología e innovación de acceso abierto, p. 2. Concytec, Lima (2013). https://portal.concytec.gob.pe/images/stories/images2013/portal/areas-institucion/dsic/ley-30035.pdf. Accessed 30 Mar 2019
6. Decreto Supremo N° 006-2015-PCM. Reglamento de la Ley N° 30035, Ley que regula el Repositorio Nacional de Ciencia, Tecnología e Innovación de Acceso Abierto, p. 4. PCM, Lima (2015). http://portal.concytec.gob.pe/images/stories/images2013/portal/areas-institucion/dsic/reglamento_repositorio_nacional_alicia.pdf. Accessed 13 Apr 2019
7. Ferreras-Fernández, T., García-Peñalvo, F.J., Merlo-Vega, A.: Open access repositories as channel of publication scientific grey literature. In: ACM International Conference Proceeding Series, pp. 419–26 (2015)
8. Paez, A.: Gray literature: an important resource in systematic reviews. J. Evid.-Based Med. **10**(3), 233–40 (2017) http://doi.wiley.com/10.1111/jebm.12266. Accessed 4 Aug 2019

9. Thelwall, M., Haustein, S., Larivière, V., Sugimoto, C.R.: Do altmetrics work? Twitter and ten other social web services. PLoS One **8**(5), e64841 (2013). https://dx.plos.org/10.1371/jou rnal.pone.0064841. Accessed 5 Oct 2019. Bornmann, L. (ed.)

10. Santana, S.: Las métricas alternativas y sus potencialidades para el profesional de la salud. Rev Médica Clínica Las Condes **29**(4), 484–490 (2018). https://linkinghub.elsevier.com/ret rieve/pii/S0716864018300804. Accessed 15 May 15 2019

11. Priem, J., Taraborelli, D., Groth, P., Neylon, C.: Altmetrics: a manifesto (2010). http://altmet rics.org/manifesto/

12. Priem, J., Groth, P., Taraborelli, D.: The altmetrics collection. PLoS One **7**(11), e48753 (2012). https://dx.plos.org/10.1371/journal.pone.0048753. Accessed 13 Jan 2020. Ouzounis, C.A. (ed.)

13. Aguillo, I.F.: Altmetrics of the open access institutional repositories: a webometrics approach. In: 23rd International Conference on Science and Technology Indicators (STI 2018). Centre for Science and Technology Studies (CWTS) (2018)

14. Palmer, L.: Altmetrics and Institutional Repositories: A Health Sciences Library Experiment. Library Publications and Presentations (2013). https://escholarship.umassmed.edu/lib_articl es/142. Accessed 13 Jan 2020

15. Eléspuru, L., Huaroto, L:. Los repositorios institucionales como herramientas para medir los indicadores Altmetrics: experiencia de la Universidad Peruana de Ciencias Aplicadas (UPC). In: V Congreso Internacional de Bibliotecas Universitarias 2016 (CIBU), Santiago de Chile, p. 20 (2016). http://repositorio.pucp.edu.pe/index/bitstream/handle/123456789/ 52633/repositorios_institucionales_altmetric_elespuru_huaroto.pdf?sequence=1&isAllo wed=y. Accessed 13 Apr 2019

16. Altmetrics.com. Altmetric badges – Altmetric (2019). https://www.altmetric.com/products/ altmetric-badges/. Accessed 14 Jan 2020

17. Adie, E., Roe, W.: Altmetric: enriching scholarly content with article-level discussion and metrics. Learn Publ. **26**(1), 11–17 (2013)

18. Konkiel, S., Scherer, D.: New opportunities for repositories in the age of altmetrics. Bull. Am. Soc. Inf. Sci. Technol. **39**(4), 22–26 (2013). http://doi.wiley.com/10.1002/bult.2013.172039 0408. Accessed 12 May 2019

19. Callicott, B.B., Scherer, D., Wesolek, A.: Making Institutional Repositories Work, 1rst edn, pp. 12–26. Purdue University Press, West Lafayette (2016)

20. Ortega, J.L.: Reliability and accuracy of altmetric providers: a comparison among Altmet ric.com, PlumX and Crossref Event Data. Scientometrics **116**(3), 2123–2138 (2018). https:// doi.org/10.1007/s11192-018-2838-z. Accessed 16 May 2019

21. Altmetrics. Researcher Data Access Program, p. 1 (2019). https://www.altmetric.com/res earch-access/

22. Zervas, M., Kounoudes, A., Artemi, P., Giannoulakis, S.: Next generation institutional repos itories: the case of the CUT institutional repository KTISIS. Procedia Comput. Sci. **146**, 84– 93 (2019). https://linkinghub.elsevier.com/retrieve/pii/S1877050919300882. Accessed 13 Jan 2020

23. Robinson-García, N., Torres-Salinas, D., Zahedi, Z., Costas, R.: New data, new possibilities: exploring the insides of altmetric.com. El Prof. la Inf. **23**(4), 359–366. https://recyt.fecyt.es/ index.php/EPI/article/view/epi.2014.jul.03/16959. Accessed 5 Oct 2019

24. Altmetric.com. Altmetric API documentation (2019). http://api.altmetric.com/

25. Mingers, J., Leydesdorff, L.: A review of theory and practice in scientometrics. Eur. J. Oper. Res. **246**(1), 1–19 (2015). https://doi.org/10.1016/j.ejor.2015.04.002

26. Manhique, I., Rodrigues, F., Sant'Ana, R., Casarin, H.: Indicadores altmétricos em periódicos brasileiros da Ciência da Informação. Rev Ibero-Americana Ciência da Informação **12**(2), 515–532 (2019). http://periodicos.unb.br/index.php/RICI/article/view/9156. Accessed 23 July 2019

27. Gadd, E., Rowlands, I.: How can bibliometric and altmetric suppliers improve? Messages from the end-user community. Insights UKSG J., 31 (2018). http://insights.uksg.org/articles/10.1629/uksg.437/. Accessed 23 July 2019

Proposal of Model for Curation Digital Objects of an Oncology Research Center

Josiane Mello$^{(\boxtimes)}$ and Angel Freddy Godoy Viera

Department of Information Science, Federal University of Santa Catarina, Florianópolis, Brazil
`josianemelloci@gmail.com, a.godoy@ufsc.br`

Abstract. This article presents a proposal model for curation of digital objects from an oncology research center in Santa Catarina State. This is an exploratory research with a qualitative approach, in which the technical procedure of case study was used. From the case study it was possible to know the digital objects that make up the informational context, and the actions taken of digital curation in the studied cancer research center. The basal phases that make up the proposed model were extracted from the literature. As a result of it, there is a model proposal to support the preservation, maintenance, access optimization, use, reuse, and the promotion of added value to the digital objects of the studied cancer research center.

Keywords: Digital curation · Oncology research center · Digital objects

1 Introduction

In the Health area, the preservation of digital information for future generations becomes essential given the growing emergence of new diseases, as well as the resurgence of others. As for example in the specialty of oncology, based on data published by the Institute José Alencar Gomes da Silva, also known as the INCA, it was estimated for the 2018–2019 biennium the occurrence of 600,000 new cases of cancer annually [1].

These estimates reinforce the need and urgency to propose new cancer prevention and control actions in Brazil. It requires further research to find the cause and cure of cancer, for which patient records and existing studies are excellent sources of information to support it.

Access to this digital information to ensure further research in the future will only be possible through preservationist actions.

In this projection, digital curation is inserted as a solution for the preservation of these digital objects, aiming at their access and reuse in new long-term research.

Based on this, a proposal for a digital object curation model of a cancer research center in the state of Santa Catarina was developed in this article. Its objective is to support active management, access optimization, use and reuse, and promote the value addition of the digital objects of this research center.

© ICST Institute for Computer Sciences, Social Informatics and Telecommunications Engineering 2020
Published by Springer Nature Switzerland AG 2020. All Rights Reserved
R. Mugnaini (Ed.): DIONE 2020, LNICST 319, pp. 235–249, 2020.
https://doi.org/10.1007/978-3-030-50072-6_18

1.1 Search of Problemam

With the arrival of the 21st century, the development of information and communication technologies (ICT) has brought significant changes in all segments of society.

In the Health area, for example, these changes brought innovations, such as the Telehealth Brazil Networks Program, the expansion of the implementation of the Electronic Patient Record (PEP), among other services and processes related to Information Management in the administrative sectors of these units. Services and processes which helped with culmination in the significant increase in the birth of digital information.

In this context, the problem arises in relation to Digital Preservation and Curation, because this digital collection of varied nature (textual and multimedia digital objects) needs to be organized, stored, made available for access (current and future), reused in new consultations and/or subsequent research, and as an element "[…] supporting the functions and activities of people, groups and institutions" [2].

It is generally agreed that this type of information support requires theoretical and practical knowledge to perform maintenance. It requires them due to the physical fragility and obsolescence of hardware, software, file formats (extensions), and storage media that are necessary for the correct maintenance, interpretation, and visualization of the bits that form a digital object [2–4].

The Digital Curatorship, besides being concerned with digital preservation, is concerned with the active management of digital objects throughout the life cycle, so that they remain continuously accessible, can be retrieved when necessary, and with aggregation of data value of these digital objects (in the sense of generating new sources of information and knowledge) [5–7].

From this perspective, considering the scope of health: where digital objects are of great value for the development of society and characterized by a complex scenario of diversified nature, in which there are textual and multimedia digital objects (image, video and audio), consisting of structures (structured, semi-structured and unstructured), with different metadata and interoperability standards, protected by medical confidentiality, which access is allowed by classifying the content of the information recorded, and may be restricted or free, as established by Law Access to Information, the various resolutions of the Federal Council of Medicine (CFM), the Brazilian Constitution and the Civil Code. The guiding question to be answered in this study is: How to make digital objects available from oncology Research Centers with a view to its maintenance and subsequent access and reuse in new research?

2   Digital Curation: Concepts and Definitions

Digital curation involves maintaining, preserving, and adding value to research data throughout its life cycle. Active management of this data reduces threats to their long-term value and mitigates the risks of digital obsolescence. In addition to reducing duplication of effort in research data creation, digital curation reinforces the long-term value of existing data by making it available for reuse in high quality new research [8].

Abbot [9], from a broader perspective, defines the concept of Digital curation as a set that brings together all the activities involved in data management, from planning its creation—when systems are designed—to digitization (dealing with analog materials), selecting formats and documentation, and ensuring that such data is available and suitable for future discovery and reuse. Digital curation also includes managing large data sets for everyday use, ensuring, for example, that they can be accessed, read and interpreted continuously.

Higgins [7] assures that the focus of digital curation is to manage the entire life cycle of digital materials, so that it remains continuously accessible and can be retrieved by those who need it.

Sayão and Sales [5] draw attention to value addition and data reuse. They define that digital curation ensures the sustainability of data for the future, while it gives immediate value to data for its creators and its users. Strategic, methodological resources, and technologies involved in digital curation practices facilitate persistent access to reliable digital data by improving the quality of that data. Its research context and authenticity are checked. In this way, the curation helps to ensure that this data is valid as archival records, meaning that it can be used in future as legal evidence. The use of common standards across different data sets, provided by digital curation, creates more opportunities for cross-sectional and collaborative searches. From a financial perspective, data sharing, reuse, opportunities for further analysis, and other benefits, value and protect the initial investment in data collection.

In the last decade, Digital curation has emerged as a new broad-spectrum practice and research area that dialogues with several disciplines and practitioners of varying categories. The Digital curator combines technologies and good practices of archiving, digital preservation, and reliable digital repositories with the management of scientific data. This combination gives rise to a new area of research, full of practical and theoretical gaps to be addressed, preferably oriented by a multidisciplinary approach [5, 10].

For the purposes of this study, we considered the concepts of digital curation presented by Abbot [9] and Sayão and Sales [5]. The first one for addressing the digital management and preservation of digital objects. The second one for focusing on value addition through the use and reuse of these digital objects by the community of users. It should be noted that previous studies do not give due importance to the power that the community of users has in adding value to digital objects.

3   Curation Lifecycle Model

The literature presents a range of life cycle models to systematize the application of digital curation activities, being the Digital Curation Center - DCC [8] life cycle model the most known and applied in Digital Curation projects. DCC [8] provides a digital curation life cycle model, it reflects a high-level view of the stages needed for the successful curation and data preservation process that begins at the conceptualization or data receipt stage in the repository.

Sayão and Sales [5] explain that the model proposed by the DCC is oriented towards planning curation activities in organizations or consortiums. It helps to ensure that all steps of the cycle will be fulfilled. However, it does not imply that all organizations must complete the cycle from the first stage. The operationalization of the stages will depend on the actual needs of each organization.

The key elements of the model are: data, digital objects, and databases. At the core of the curator's life cycle is data - which is any information encoded in binary format [8].

The model that can be seen in Fig. 1 presents three types of actions that should be applied during the Digital Curation process, that is, actions for the whole life cycle. Those include description and representation of information, preservation planning, participation and monitoring, and curation and preservation. Sequential actions include conceptualization, creation and/or receipt, evaluation and selection, archiving, preservation, storage, access, use and reuse, and transformation actions. Finally, occasional actions include elimination, reevaluation, and migration. The model designed by DCC provides a collective view of the set of functions required for curation and data preservation. In addition to defining roles, responsibilities, and concepts, it introduces the standardization infrastructure and technologies that must be implemented [5].

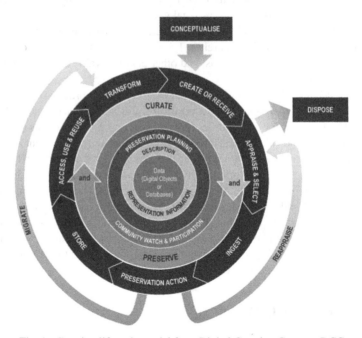

Fig. 1. Curation lifecycle model from Digital Curation Center – DCC.

4 Digital Objects in Health

For this proposal, it will be considered the concept of digital object proposed by Yamaoka and Gauthier [11] based on Ludwig [12], due to its broad scope of conceptual scope - as shown in the concept map expressed in Fig. 2.

Thus, digital object means any information object of any type and format expressed in digital form. Which, as Thibodeau [13] assures, inherits properties of three layers:

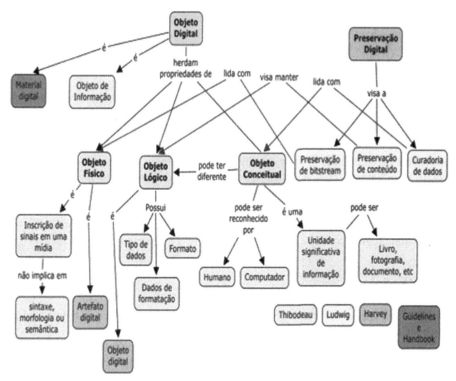

Fig. 2. Conceptual map of a digital object based on the model of Thibodeau [13] and other authors. **Source:** Yamaoka and Gauthier [11] based on Ludwig [12].

physical object, logical object, and conceptual object. The digital object deals with the physical object and the conceptual object. It aims to maintain the logical object.

In the physical object layer, the digital object is simply an inscription of signals in a midst. Not implying syntax, morphology, or semantics. The physical object is also called a digital artifact by Harvey [14]. At the logical object layer, the digital object is recognized and processed by software. Grammar does not interfere, and the format, formatting data, and code (ASCII) data types of the object are recognized by application software.

Harvey [14] calls the logical object a digital object. In the conceptual object layer, the digital object can be recognized and understood by humans or computers. It is a significant unit of information. It can be a book, a map, a photograph, a memo, and so on. Coding in the conceptual object may differ from the logical object, and this has implications for its preservation. Digital preservation aims at bitstream preservation, content preservation, and data curation [12].

Based on the case study conducted at a Santa Catarina cancer research center, the following digital objects in health were identified as:

– Electronic Patient Record (PEP): This is a unique document, consisting of textual properties (identification forms, prescriptions, examination requests, etc.) and multimedia (examinations in image and video format), multifunctional (generated and used for various purposes) and multi professional (registered and consulted by a multidisciplinary team) [15, 16].
– Primary studies: These are medical studies that correspond to original investigations. They constitute the majority of publications found in medical journals and are classified as: Case report; Case Studies and Controls, Case Detection - Screening, Cohort Study, and Randomized Controlled Trial [17].
– Secondary studies: seek to draw conclusions from primary studies with a brief record of findings that are common to them. These studies correspond to: Reviews, Systematic Reviews, Meta-Analysis, Guidelines, Decision Analysis, and Economic Analysis [17].
– Scientific publications: include articles, thesis, dissertations, and reports (estimates of new cancer cases, tumors with higher incidence rate, distribution of incidence by geographic region, and deaths by cancer types for a given period).

5   Methodology

This is an exploratory research with a qualitative approach, applied in nature and in which the technical case study procedure was used.

This case study was conducted at a Santa Catarina cancer research center in April 2019, after being approved by the ethics committee and signed by its participants, in a Clear and Informed Consent Form (ICF).

It aimed to know the digital objects and their characteristics, and finally the digital curation actions undertaken in this place.

The concept of digital object adopted in this research was proposed by Yamaoka and Gauthier [11], as described in the literature review, in the section that deals with digital objects. The choice for this concept is justified by the broad conceptual scope.

As for the digital objects belonging to the center studied, the research identified the following: Patient's electronic records, Primary studies, secondary studies and the scientific publications.

Regarding the actions taken at the center studied regarding digital curation, the research found the use of backups, the implementation of the electronic medical record via TASY system and the use of PACS software for automated treatment of image and video exams.

It should be noted that these initiatives do not ensure the maintenance, long preservation and value addition of these digital objects.

In this sense, a model proposal for digital curation of digital objects is presented, based on the Digital Curation Center (DCC) life cycle model adapted to the reality of the cancer research center studied, aiming at preservation, maintenance and value aggregation of these digital objects.

6   Model Proposal for Curating Digital Objects from an Oncology Research Center at Santa Catarina

Based on the theoretical framework presented, this paper presents a proposal for a generic model to curate digital objects from cancer research centers. This proposal will be developed based on the life cycle model proposed by the Digital Curation Center [8]. It is aimed to ensure the active management, preservation, and value addition of digital objects from cancer research centers.

To do so, the basic phases that constitute the present proposal will be presented. Figure 3 presents a macro view of these phases, followed by their description.

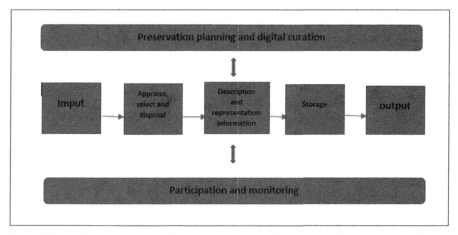

Fig. 3. Flow of the macro phases that make up the model proposal. **Source:** Prepared by the authors.

As already mentioned, this proposal aims to ensure the active management, preservation and value addition of the following digital objects of the studied center: electronic patient record (PEP), primary and secondary studies, and scientific publications.

Therefore, guidelines are proposed regarding the implementation of a curated project for these types of digital objects: textual and multimedia, structured, semi-structured, and unstructured. They have different metadata and interoperability standards. They are protected by medical confidentiality, whose access is allowed, based on a set of laws.

This model proposal has the function of systematizing the digital curation process, assisting the curator in a practical way in the administrative processes and management of digital objects, as well as in the selection. In the selection he will decide for the preservation, or not, of a certain object and its digital transformation (giving rise to new digital objects from the original).

It should be clarified that the following phases must be employed and strictly adhered to in order to ensure the authenticity, reliability, integrity, and usability of the digital object [19].

a) Input

In the model proposed for cancer research centers, the curator may be an IT professional, librarian, or archivist. Digital objects are understood as any information object of any typology and format expressed in digital form that inherits properties of the object layers: physical, logical and conceptual [7–9, 18].

The digital objects dealt here are: electronic patient record (PEP), primary and secondary studies, and scientific publications.

In the input phase, the curator conceptualizes the digital object received or created in order to identify what type of object it is. The professional designs and plans digital object creation. It includes capture methods and storage options. Issues such as intellectual property, embargoes and restrictions, funding, responsibilities, specific research objectives, capture and calibration tools should be recorded [5].

It should be noted that the receipt will happen in accordance with the policy, rules, administrative acts, and other regulations established by the maintaining center. This information is recorded to characterize authorship (intellectual property), typology, access, embargoes and restrictions, funding, responsibilities, specific objectives, appropriate metadata for curation and preservation, capture tools, and calibration.

After the conceptualization, whether the digital object is received or created, the curator prepares the digital object to enter the CD cycle. The curator assigns to it administrative, descriptive, structural, technical, and preservation metadata.

In the case of self-archiving, the author plays the role of curator and inserts the metadata - obeying the policies, rules, administrative acts, and other regulations established by the maintaining center.

The input phase steps are shown in Fig. 4.

b) Appraise, selection and disposal

Once conceptualized, the curator evaluates the digital object received applying sound selection criteria set out in established policies, guidelines or legal requirements by the cancer research center. Based on these criteria, the curator decides whether the object will be selected or not for long term preservation. Digital objects that have not been selected for long-term preservation will be discarded or relocated to a file or other custodian. In some cases, depending on the nature of the digital object, legislation may indicate safe destruction (Fig. 5).

c) Description and representation of information

Once selected the digital objects that will be part of the curation process, they will receive appropriate technical treatment. It will be assigned administrative, technical, structural, and preservation metadata according to the appropriate standards. In the research, it was found that the chosen center uses the following metadata and interoperability standards: ICD 10, SUS Table, DICOM Standard, HL7, openEHR, EAD, Dublin Core and Z39.87.

The description and thematic representation of the information contained in the digital object can be done through the AACR2 (Anglo-American Cataloguing Rules), RDA (Resource Description and Access), DDC (Dewey Decimal Classification) or UDC (Universal Decimal Classification) - based on the policies of the digital objects holding units of the studied center. In the case of electronic medical records, the technical treatment of the data is done using the tools provided by the Philips Tasy system itself. The same

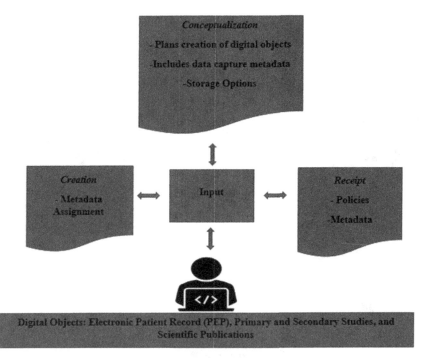

Fig. 4. Input phase steps. **Source:** Prepared by the authors.

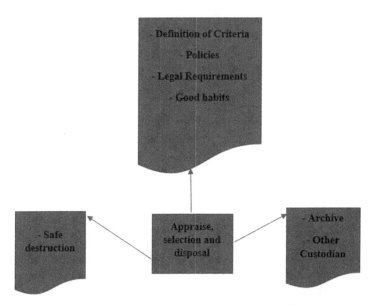

Fig. 5. Appraise, selection and disposal. **Source:** Prepared by the authors.

is applied for imaging and video examinations through the Communication and Image Archiving System (PACS) system by the radiology sector (Fig. 6).

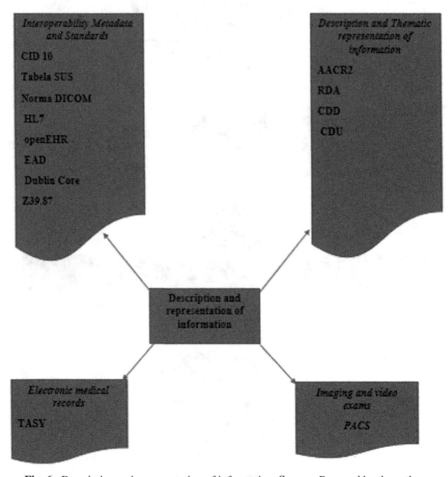

Fig. 6. Description and representation of information. **Source:** Prepared by the authors.

d) Storage

At the end of the technical processing phase of digital objects, they will be ready to be stored in databases, institutional repositories, and other digital platforms to be found by users in need - which in this case are: doctors, nurses, nursing technicians, pharmacists, nutritionists, social workers, cancer patients, family members, and other employees of the study center. They will be able to be found in the present and in the future, regardless of the format or current technology.

Storage in repositories should be based on preservation and access policies. Preservation in these environments can be accomplished through backup, metadata use, and preservation strategies: migration, emulation, refresh, encapsulation, and technology preservation, and adherence to the requirements set out in the Open Archival Information System (OAIS) protocol reference model. The OAIS reference model became in 2003 an international standard, ISO 14721: 2003, with the objective of establishing an archiving system information through a schema order to preserve these and make them available to a designated community [20].

Elements must be stored in preservation formats in storages with their metadata. Preservation storage should be performed on a separate infrastructure from web servers for user access as shown in Table 1 [20].

Table 1. Definitions of file formats for preservation and access. Source: Siebra et al. [20].

Tipo	Formato	Base
Portable Document Format Archives (PDF/A)		
Texto (preservação)	PDF/A-1	ISO 19005-1:2005
	PDF/A-2	ISO 19005-2:2011
	PDF/A-3	ISO 19005-3:2012
Texto (acesso)	Portable Document Format (PDF)	Adobe Systems Incorporated
Imagem (preservação)	Tagged Image File Format (TIFF)	Adobe Systems Incorporated
	TIFF, Revision 6.0 and earlier	
	TIFF Uncompressed File with Exif Metadata	
Imagem (acesso)	JPEG file with Exif metadata	ISO/IEC 10918
		ISO/IEC 14495
Áudio (preservação)	Broadcast WAVE file, version 1, with LPCM encoded audio	EBU Tech 3285 - Specification of the Broadcast Wave Format (BWF) - Version 1 - second edition (2001)
	Broadcast WAVE file, version 2, with LPCM encoded audio	
Áudio (acesso)	MP3 (MPEG Layer III audio encoding)	MPEG-1: 11172-3
		MPEG-2: 13818-3
Video (preservação)	MP4 File Format	ISO/IEC 14496-14:2003
	MPEG-4 file format, version 2 (sem compactação)	
Video (acesso)	MPEG-4 com compactação	ISO/IEC 14496-2:2004

e) Output

At this stage, it deals with issues related to access, use and reuse of digital objects. In this sense, products and services are developed to promote digital objects submitted to the curation process. It includes: repositories, databases, digital object access products, search and retrieval services. In the design of these digital informational environments, it is recommended to consider the main attributes of Information Findability (AEI) established by Vechiato [21], namely: Navigational Taxonomies, Folksonomies, Computer Mediation, Affordances, Information Discovery, Accessibility and Usability, Mediation, Mediation of subjects, Intentionality, Mobility, Convergence, and Ubiquity. They are recommended in order to optimize access and use of information. Still in this phase, it includes the participation of the user community. They can access through the internet the contents of digital objects already archived, produce new content from them, as well as provide information about a certain digital object - aiming to enrich this resource through its context and its domain information. The platform will record community interactions with digital objects and the effects of this interaction. This registration can help to adapt the way resources are made available to users, as well as the arrangement of objects in the repository or system through the principles of Information Architecture. It is possible to use ontologies to link a document to related ones through their entities. Its purpose is to reference other objects of the same subject, and, thus, keep the user informed of the interrelationship of these objects in order to indicate the contributions and projection for new research. All this interaction of users will be recorded in the system, accessed, used, and reused in order to contribute to the process of planning their digital preservation. It can be considered the transformation or reevaluation of a digital object.

It is noteworthy that their access, use, and reuse will occur through rules of access restrictions, copyright, and rules for data dissemination and sharing. It should be noted that digital objects from an oncology research center carry a range of sensitive information protected by medical confidentiality. It has restricted access by others, as set forth in the law of access to information, in several resolutions of the Federal Council of Medicine (CFM), in the Brazilian constitution and in the Brazilian civil code.

f) Participation and monitoring

In this phase the user community is assisted, as well as the activities that take place under them. All interaction will be mapped through statistical reports to investigate if reuse of "cured" digital objects is happening and at what level it is happening. That is, it will be identified if cured digital objects are resulting in new research, new procedures and the impact of this on cancer treatment (based on data presented in previous years reports on the evolution of this disease).

At this phase, is also noticed the evolution of technological tools, social media, and applications. Also, it is noticed how they can contribute to leverage the process of adding value to cured digital objects of the center (Fig. 7).

g) Preservation planning and digital curation

This phase includes activities that permeate the entire digital curation lifecycle on an ongoing basis. At this stage, the physical structure necessary to support such a project is dimensioned, from the present moment to a temporal projection date. Here it is analyzed storage data servers, servers, and other physical components needed to achieve the desired goal in this project. An analysis of the archives typology is performed and the formats for their access and their preservation are defined, as shown in Table 1. Based on this, the preservation strategies are timely, security measures are established, and finally,

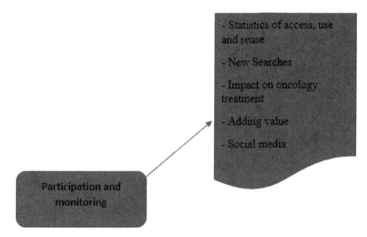

Fig. 7. Participation and monitoring. **Source:** Prepared by the authors.

people are assigned responsible for the custody and actions that ensure the full exercise of the digital curation process for the digital objects of the research center. For each of these planning steps, a checklist will be developed as a quality tool to ensure that tasks are performed as scheduled.

This phase also focuses on transformation and reevaluation activities. Based on the access and use and reuse of digital objects, a certain digital object is forwarded to the selection and evaluation stage. It is forwarded for further consideration and decision as to whether discard it or not. In other cases, it may be recommended to transform the digital object, that is, create others from the original to meet new demands.

7   Conclusion

The development of this model proposal for the digital curation of digital objects of an oncology research center was motivated thinking about allowing the access, use and reuse of information from the center afterwards. It is believed that the development of this framework will support health researchers in tracking, controlling, and decreasing the number of new cancer cases. That said, as it is a proposal, it is recommended as future work, the application of this in a unit of cancer research to investigate their adherence to the processes of preservation, maintenance, and value addition of digital objects. Also, focusing on investigate the impact there of figures presented in the INCA reports on cancer in the coming years.

References

1. Instituto Nacional de Câncer José Alencar Gomes da Silva - INCA. Coordenação de Prevenção e Vigilância. Estimativa 2018: incidência de câncer no Brasil. Instituto Nacional de Câncer José Alencar Gomes da Silva. Coordenação de Prevenção e Vigilância. INCA, Rio de Janeiro (2017). http://www1.inca.gov.br/inca/Arquivos/estimativa-2018.pdf. Acesso 03 ago 2019

2. Márdero Arellano, M.Á.: Critérios para a preservação digital da informação científica. 2008. 354f. Tese (Doutorado em Ciência da Informação) - Universidade Federal de Brasília, Departamento de Ciência da Informação (2008). http://repositorio.unb.br/bitstream/10482/1518/1/2008_MiguelAngelMarderoArellano.pdf. Acesso 15 mar 2016

3. Beagrie, N., Jones, M.: Preservation Management of Digital Materials: A Handbook. British Library, London (2001)

4. Santos, H.M., Flores, D.: Os impactos da obsolescência tecnológica frente à preservaçao de documentos digitais. Braz. J. Inf. Sci. **11**(2) (2017). http://www.brapci.ufpr.br/brapci/v/a/23322. Acesso 19 out 2017

5. Sayão, L.F., Sales, L.F.: Curadoria Digital: um novo patamar para preservação de dados digitais de pesquisa. Informação e Sociedade: Estudos, João Pessoa **22**(3), 179–191 (2012)

6. Lee, C.; Tibbo, H.: Digital curation and trusted respositories: steps toward success. J. Digit. Inf. **8**(2) (2007). http://journals.tdl.org/jodi/. Acesso 20 dez 2017

7. Higgins, S.: Digital curation: the emergence of a new discipline. Int. J. Digit. Curation **6**(2), 78–88 (2011)

8. Digital Curation Centre – DCC: What is Digital Curation? University of Edinburgh, Edinburgh (c2004-2016a). http://www.dcc.ac.uk/digital-curation/what-digital-curation. Acesso 10 set 2019

9. Abbot, D.: What is digital curation? Digital Curation Centre, Edinburgh, UK (2008). http://www.dcc.ac.uk/resources/briefing-papers/introduction-curation/what-digital-curation. Acesso 02 set 2019

10. Dutra, M.L., Macedo, D.D.J.: Curadoria digital: proposta de um modelo para curadoria digital em ambientes big data baseado numa abordagem semi-automática para a seleção de objetos digitais. Informação Informação **21**(2), 143–169 (2016). http://www.brapci.inf.br/v/a/22209. Acesso 04 set 2019

11. Yamaoka, E.J., Gauthier, F.A.O.: Objetos digitais: em busca da precisão conceitual; objetos digitales: en busca de la precisión conceptual. Informação Informação **18**(2), (2013). http://www.brapci.inf.br/v/a/13516. Acesso 20 set 2019

12. Ludwig, J.: About the Complexity of a Digital Preservation Theory and Different Types of Complex Digital Objects. Schloss Dagstuhl - Leibniz-Zentrum fuer Informatik, Dagstuhl, Germany (2010). http://www.dagstuhl.de/Materials/Files/10/10291/10291.LudwigJens.Ext Abstract.pdf. Acesso 20 set 2019

13. Thibodeau, K.: Overview of Technological Approaches to Digital Preservation and Challenges in Coming Years. The State of Digital Preservation: An International Perspective. Anais…CLIR and Library of Congress, Washington (2002)

14. Harvey, R.: Preserving Digital Materials. Saur Verlag, München (2005)

15. Pinto, V.N.B.: Prontuário eletrônico do paciente: documento técnico de informação e comunicação do domínio da saúde. Encontros Bibli: Revista Eletrônica de Biblioteconomia e Ciência da Informação **11**(21), 34–48 (2006). http://www.brapci.ufpr.br/brapci/v/a/3837. Acesso 20 out 2017

16. Lunardelli, R.S.A., Tonello, I.N.M.S., Kawakami, T.T.: O prontuário eletrônico do paciente no contexto nacional: resultados de um projeto de pesquisa. In: ENCONTRO NACIONAL DE PESQUISA EM CIÊNCIA DA INFORMAÇÃO, 16, 2015, João Pessoa, PB. Anais eletrônicos… João Pessoa, PB (2015). http://www.brapci.inf.br/index.php/article/download/43941. Acesso 16 set 2019

17. Campana, A.O.: Metodologia da investigação científica aplicada à área biomédica – 2. Investigações na área médica. J. Pneumol. **25**(2) (1999). http://www.scielo.br/pdf/jpneu/v25n2/v25n2a5.pdf. Acesso 19 set 2019

18. Yamaoka, E.J.: Ontologia para mapeamento da dependência tecnológica de objetos digitais no contexto da curadoria e preservação digital. AtoZ: Novas Práticas em Informação e Conhecimento **1**(2), 65–78 (2012). https://doi.org/10.5380/atoz.v1i2.41313. Acesso 10 out 2019

19. Pennock, M.: Digital curation: a lifecycle approach to managing and preserving usable digital information. Libr. Arch. [S.l.] **1**(18), 1–3 (2007). http://www.ukoln.ac.uk/ukoln/staff/m.pen nock/publications/docs/lib-arch_curation.pdf. Acesso 18 set 2019

20. Siebra, S.A., et al.: Projetos de curadoria digital: um relato de experiências. Bibliotecas. Anales de Investigación (Cuba) **14**(2), 164–178 (2018). http://hdl.handle.net/20.500.11959/brapci/60013. Acesso 11 out 2019

21. Vechiato, F.L.: Encontrabilidade da informação: contributo para uma conceituação no campo da Ciência da Informação. (2013). 206 f. Tese (Doutorado em Ciência da Informação) - Faculdade de Filosofia e Ciências de Marília. Universidade Estadual Paulista (2013)

Author Index

Printed in the United States
By Bookmasters